D0570240

the GIS 20 essential skills

third edition

GINA CLEMMER

Esri Press
REDLANDS | CALIFORNIA

Esri Press, 380 New York Street, Redlands, California 92373-8100
 Third edition 2017
Printed in the United States of America
21 20 19 18 17   1 2 3 4 5 6 7 8 9 10

Library of Congress Cataloging-in-Publication Data

Names: Clemmer, Gina, 1974- author.
Title: The GIS 20 : essential skills / Gina Clemmer.
Other titles: The Geographic Information Systems Twenty
Description: Third edition. | Redlands, California : Esri Press, [2018] |
Includes index.
Identifiers: LCCN 2017043415 (print) | LCCN 2017053155 (ebook) | ISBN
9781589485136 (electronic book) | ISBN 9781589485129 (paperback : alk.
paper)
Subjects: LCSH: Geographic information systems. | ArcGIS.
Classification: LCC G70.212 (ebook) | LCC G70.212 .C58 2018 (print) | DDC
910.285--dc23
LC record available at https://urldefense.proofpoint.com/v2/url?u=https-3A__lccn.loc.gov_2017043415&d=DwIFAg&c=n6-cguzQvX_tUIrZOS_4Og
&r=RhmcbAxStnbJpr06ef1onNDeVX-gjVopdqeQ8i7DbIY&m=GJvL4rmn8BCmwiDDWsWOZNpqfrW4lbbRhalvgz_G-2E&s=mH7SASjq_tnRvenGLju_
hG8FJ00OhJMaV83PNeRaNGs&e=

Ask for Esri Press titles at your local bookstore or order by calling 800-447-9778, or shop online at esri.com/esripress. Outside the United States, contact your local Esri distributor or shop online at eurospanbookstore.com/esri.

Esri Press titles are distributed to the trade by the following:

In North America:
Ingram Publisher Services
Toll-free telephone: 800-648-3104
Toll-free fax: 800-838-1149
E-mail: customerservice@ingrampublisherservices.com

In the United Kingdom, Europe, Middle East and Africa, Asia, and Australia:
Eurospan Group
3 Henrietta Street
London WC2E 8LU
United Kingdom
Telephone: 44(0) 1767 604972
Fax: 44(0) 1767 601640
E-mail: eurospan@turpin-distribution.com

Contents

A note to the reader	vii	
Introduction	ix	
Chapter	*Page*	*Skill*
1	1	Downloading shapefiles and using essential ArcMap tools
2	17	Creating basic maps and layouts
3	31	Projecting shapefiles
4	39	Preparing data for ArcMap
5	51	Joining data to maps
6	57	Creating thematic maps
7	67	Working with data tables
8	75	Address mapping
9	91	Creating a categorical map
10	95	GPS point mapping
11	99	Editing
12	109	Creating attribute queries
13	113	Creating location queries
14	117	Using geoprocessing tools
15	127	Creating geodatabases
16	133	Joining boundaries
17	137	Working with aerial photography
18	145	Creating reports
19	151	Sharing work
20	159	Publishing maps
Thanks	169	
Source credits	171	
Index	175	

A note to the reader

Hey, reader. This book is an extension of my passion to help everyday people quickly learn the fundamentals of ArcGIS®. I have taught students 16 to 85 years old from all walks of life—fishing villages in Alaska, executives in Times Square, bureaucrats in Washington, DC. And with slight variation, they all wanted to do the same key things with ArcGIS. The three main requests are reference mapping, thematic mapping, and geocoding. It is my hope to have provided you with 20 easy-to-follow exercises that illustrate these principles.

Over the past decade, it has been my passion to provide an applied approach to teaching geographic information systems (GIS) to busy professionals. I believe the fundamentals of GIS (and ArcMap™, specifically) can be taught in the following ways:

- quickly. I do not believe GIS is the deeply complicated discipline many make it out to be.
- using a project-based paradigm in which completing a concrete task is the goal versus using a layered or building-block approach. In other words, I don't think you must know the inner workings of the GIS to create common types of maps.
- by learning what is most frequently used by most people (and therefore most important to understand) and skipping the rest. I do not believe you must know every aspect of ArcMap to successfully complete common GIS tasks.
- using everyday language versus technical jargon.

The 20 chapters that follow reflect 20 years of market research from 20,000 students asking me "How do I ...?" These students were just like you: new to GIS, some excited about the prospect of being able to map their data, some only there because their bosses said they had to be—but after learning the fundamentals of ArcGIS, all were GIS converts.

Once you see the power of GIS, and see that you yourself can create maps easily, you simply can't unsee it. GIS will change the way you think about presenting data and solving problems. It is my hope to give you ideas and tools, and through sophisticated problem-solving and the persuasion of maps, you can go out there and do good work and make the world a better place.

Gina Clemmer
Portland, Oregon
September 2017

Introduction

This section has some superimportant stuff in it, including how to get the software, what ArcMap is, and a few tricks of the trade.

Downloading and installing ArcGIS® Desktop software (180-day use)

If you have the software, skip this part.

You must have an installed copy of ArcGIS Desktop to complete these exercises. The book was updated for ArcGIS® Desktop 10.6. It also works for 10.5 and related service packs and patches. But any license level (Basic, Standard, or Advanced) and most any version will work. The geocoding chapter (chapter 8) is contingent on version 10.1 and higher.

You can download the ArcGIS Desktop Advanced 180-Day Trial for Esri Press with the purchase of this book.

NOTE: *Esri Press e-books do not include 180-day trial software.*

The 180 days begins when you install and register the software. It is a good idea not to register the software until you are ready to use it. The software will become inactive 180 days after registration, regardless of whether you have ever used it.

Important stuff to know

- If you have a previous version of the software on your computer, you must first uninstall that software before installing a new version. Beware, the installation takes time.
- The software works only on PCs using Windows® 8.1 and higher.
- You must have Internet Explorer® installed on your computer (although you do not need to use it for the download). Other parts of ArcGIS use Explorer.

Download the software

1. **Go to the book resource web page at esri.com/GIS20-3, an d click the link for the software trial.**

2. **Near the upper-left corner of the page, click Create an Account tab. You must establish an Esri Global Account before you can download the software. Enter your first and last name and email address. Then click Create an Account.**

3. **Go to your email and find an instant email from Esri®. If you do not see it immediately, check your spam folder. Click the Activate your Account button in the email.**

4. **On the account page, you must input a user name and password.** At this point, you're most likely to think, "Wait, didn't I just do that?" But you didn't really. Keep going.

5. **Fill in all the gaps, select the check box to accept the terms and services, and then click Create.**

   TIP ***Write down this user name and password, or put it in your phone. The user name doesn't have to be from an email, but it's a good idea to make the user name your email address so it's easier to remember.***

   After you click Create, a thank-you message appears, but nothing happens with the download you were trying to download. So now you must go back to the link.

6. **On the link page, input the user name and password you just created, and click Sign In.** On the inside back cover of your book, there's a stamped 12-character authorization number (EVAxxxxxxxxx). You'll need this number for the 180-day trial. Esri Press e-books do not come with the 180-day trial.

   NOTE: *The code works for only one download.*

7. **Scroll down to the bottom of the page, and click Download for ArcGIS Desktop. (Note: If it doesn't begin downloading immediately, look in the upper-right corner of your browser, and make sure you are not getting an alert that you must enable pop-ups. If you are, click the alert, and click the option to always allow pop-ups from** esri.com.**)**

8. **Click "Click to download your file now." If you get an alert that says this type of file may harm your computer, you must approve the download. If you're using Google® Chrome™ as a browser, click Keep to start the download.** Depending on your internet connection speed, it may take 10 minutes or more to download.

Install the software

1. **After the file finishes downloading, click on the downloaded .exe file.** A pop-up window for Destination Folder for Files will appear.

2. **Generally, just click OK unless you prefer to store the files at a location other than your C drive.** The .exe file will unzip.

3. **Once the files are unzipped, click Close, and the installation program will begin to install. You may get a pop-up that asks if it's okay to allow this app to make changes. Click Yes.** The application will begin to "unpack," which is the same thing as unzipping.

4. **Click Next, and accept the terms of service when prompted. For this complete install, continue clicking Next through the menus when prompted.** Get a cool, refreshing beverage. It's going to take some time.

Authorize your trial

1. **Before you can use the software, you must input your authorization number, which acts as a temporary license. Retrieve the email from Esri with your evaluation authorization number.** The number will begin with EVA.

2. **Next, go to the Start menu, and click ArcGIS in your list of programs. You'll see a submenu, which has lots of entries. Click ArcGIS Administrator. If prompted, allow the program to make changes.**

You're going to need data to complete these exercises

Go to esri.com/GIS20-3 to get the chapter data files.

Other good stuff to know

Language

The GIS 20 explains ideas and steps in nontechnical, everyday language using terms that the average person can understand, not GIS jargon. These explanations are particularly helpful for new users of ArcGIS and people who speak English as a second language. Following are some common points of confusion.

What is the difference between ArcGIS and ArcMap?

ArcGIS and ArcMap are often used interchangeably. This book uses the terminology "ArcMap" during the exercises, because it is short and easy to reference. ArcMap is actually the software's browser; it is not the whole software (just as Internet Explorer is an internet browser, not the internet itself).

ArcGIS Desktop refers to the entire mapping software suite available for use on a PC. The software has gone from ArcView to ArcGIS to ArcGIS Desktop to distinguish it from ArcGIS for other platforms, such as ArcGIS® Pro and ArcGIS® Enterprise.

The exercises in this book can be done using the most basic version of the software, which is ArcGIS® Desktop Basic, Single Use. Other licenses give access to other extensions used for specialized tasks, but these extensions are not used in this book.

What is the difference between shapefiles, layers, and .shp files?

In everyday language, shapefiles are called many different things: layer files, boundary files, geography files, and, of course, shapefiles. This book uses the terms "shapefile" and "layer" interchangeably, but do not confuse this use of the word "layer" with the actual layer file type (.lyr).

Layer files are mentioned in chapter 20. When they are used, we are clearly discussing a layer file type, not a shapefile. Assume every file is a shapefile unless it is specifically identified as a different file type.

Also, sometimes the full name of the shapefile is used, such as "states.shp," but then later in the text, the name is shortened to "states," or it is referred to generally as the "states layer" or "states shapefile." All of this refers to the same file, so try not to get confused.

Tricks of the trade

This section covers those things that someone who has been working with ArcMap for a long time just knows. Every profession has them—tricks of the trade—things that will make your chances of success better. The "tricks" covered here are things you really must know to be happy and successful with this software.

Finding files—here's the secret

Steps for finding files are not repeated in each chapter of the book. Who wants to repeat themselves constantly? So here is what you need to know. Read the "connect to folder" discussion in chapter 1 carefully. You must first "connect to folder" to locate and connect to your files. Novices who have not read the information you are reading now spend a tremendous amount of time and frustration trying to find their files. Additionally, various file types are visible in different places, adding to the confusion. For example, shapefiles (.shp) cannot be viewed from the Open menu, and ArcMap documents (.mxd) cannot be viewed when using the Add Data tool.

If you become comfortable navigating ArcMap and saving and opening your files, not just kind of comfortable, or sort of comfortable, but very comfortable doing this, you are guaranteed to learn this software more quickly and with less frustration.

'Right-click > Properties'—the answer to every question

If one tip can make your life easier, it is to right-click the layer name in the table of contents, and then click Properties. Shorthand such as "right-click > Properties" is often used to describe this function. Right-clicking a layer and clicking Properties will give you access to all available options for that layer.

This right-click > Properties trick also works with almost all items in ArcGIS Desktop. "Right-click > Properties" is the answer to almost every "How do I ...?" question in ArcMap. If you are ever in doubt as to what to do, click the item to make it active, and then right-click > Properties. This step will generally get you in the right neighborhood.

Toolbars

This software comes with 46 toolbars encompassing hundreds of tools. You can access toolbars in ArcMap by going to the Customize menu and clicking Toolbars. You'll find three toolbars indispensable: Standard, Tools, and Draw.

It is a good idea to enable these toolbars before starting your work. Once they are enabled, they stay enabled until you turn them off. If there are certain sets of tools you use repeatedly, you can create customized toolbars.

Help menu

Because this book does not cover every aspect of ArcGIS Desktop, learning to use the help menu will be, well, helpful. Two help menus are built into the ArcMap interface: ArcGIS Desktop Help, which comes with the software, and ArcGIS Desktop Web Help (formerly called the ArcGIS Resource Center). The web help is frequently updated and loaded with new information. Both are accessible from the ArcMap window under Help (upper right).

Internet browsers

The official recommendation for ArcGIS Desktop-compatible internet browsers is Internet Explorer 8.0 and higher. Geocoding will not work without Explorer on your computer. With this exception, any browser will work for these exercises.

CHAPTER 1

KEY CONCEPTS
downloading shapefiles
learning essential tools
exploring the ArcMap table of contents
understanding shapefiles
customizing shapefiles
saving projects

Downloading shapefiles and using essential ArcMap tools

Shapefiles are the building blocks of many maps and a natural place to begin. Shapefiles are layers that can be stacked on top of each other, like a layered cake, to create one composite map image. Shapefiles contain both the map and the underlying data for the map. In your GIS life, and in this book, you will need various types of shapefiles. You will also need an understanding of essential tools and key ArcMap features such as the table of contents.

Anyone who watches cooking shows knows that they breeze right over all the prep work—well, that's not the case here. In this exercise, you'll download shapefiles and learn key ArcMap tools.

[*You will need to download all chapter data files from esri.com/GIS20-3.*

Downloading shapefiles

Shapefiles are the maps behind the maps. You need them, and you need to know a few things about them. You need to know how fresh the shapefile is (is it from 2016 or 1972?), how much it costs to get your hands on it (free or $2,000? Makes a difference), and how accurate it is (drawn by fifth-graders or NASA?). The US Census Bureau is an outstanding resource for shapefiles. It updates them every year, they cost zero dollars, and they are pretty accurate.

The US Census Bureau is the national custodian of geographic definitions of US borders. It is the agency in charge of keeping the official latitude-longitude boundaries of all states and the US as a whole.

In the following chapters, and in life in general, you will need shapefiles to create maps. Two useful shapefiles are counties and cities. You can download these two now. But first, you will need a place to save your files.

Set up a save folder

1. **Using Windows Explorer (outside of ArcMap), create a folder on your C drive, such as C:\GIS20, where you will save files for this exercise and others.** If you don't know how to create a folder on your C drive to store files, you should ask someone. Going forward, this folder will be referred to as your "save folder." It's important to save on your root C drive because performing the simple task of finding your files from within ArcMap is not as easy as you might think.

Select files from the US Census Bureau website

1. **Navigate to the US Census Bureau website, https://www.census.gov.**

2. **On the Census Bureau site, at the top of the page, click the Geography tab, and then click the Maps & Data button. Then in the right column under Geographic Data, click the TIGER Products link.**

3. **Click TIGER/Line® Shapefiles - New 2016 Shapefiles link. (These files are updated annually, so click the most current year.)**

4. **Scroll to the bottom half of the page, and click the 2016 tab.**

5. **Click the Download expandable section header, and then click the Web interface link.** You've just entered the secret portal where current, free shapefiles live.

Download county, city, and state shapefiles

First, you must select what type of geographic file to download.

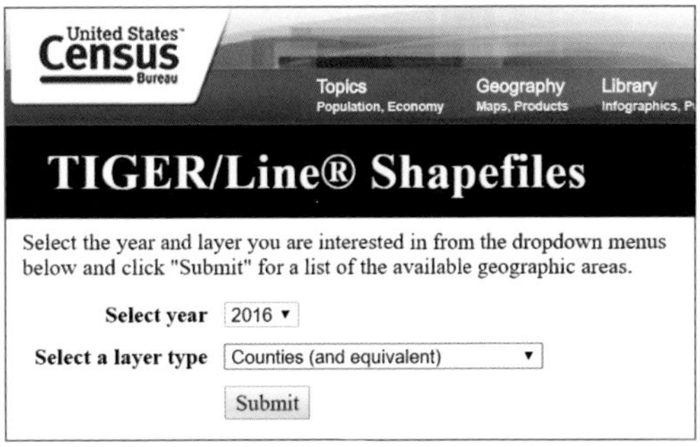

1. **From the "Select a layer type" list, click Counties (and equivalent), and then click Submit.**

2. **On the next page, click Download national file.** The County shapefile will begin to download. This file includes all counties in the US. One way or another, your mission is to get these zipped files into your save folder so they can be unzipped and you can find them again when the time comes.

   In Internet Explorer, you will be asked whether you want to Open, Save, or Cancel the file.

3. **Click Save to save the zipped file to your C:\GIS20 folder.**

4. **On the Census site, after the file downloads, click the back arrow once to return to the "Select a layer type" list.**

5. **From the drop-down list, select Places, and then click Submit.** "Places" is what the Census Bureau calls cities. It includes all US cities, as well as something called Census Designated Places (CDPs), which are not cities per se but highly recognized areas. The thing to know here, though, is that this file is *the* city file.

6. **Select the state in which you live (you'll use this state in many subsequent exercises), and then click Download.** In this example, Alabama will be selected. Why Alabama? Because it's first on the list. You can pick your own state.

   Optional: you might want to also grab the State (and equivalent) shapefile, too.

   If you used Internet Explorer to download files, they should already be in your save folder. If you used Chrome, you will move the files from the Downloads folder to your save folder.

7. **Access the Downloads folder by clicking the Chrome settings in the upper-right corner, and look for three little dots stacked on top of each other. Click that button and then Downloads, and then click the "Show in folder" link. (You'll see the zipped file there.) Right-click once on one of the files, click Cut, navigate to your save folder, and right-click Paste (or drag and drop if that's possible).**

8. **Unzip the files (most PCs come with a built-in "unzipper"). Try right-clicking on the zipped file, then clicking Extract All, and, here's where people get really confused, you must click the Browse button , navigate to your save folder, and save there.** Unfortunately, there's no way to unzip all the files all at once. You must unzip each file, one at a time.

   If you are at all confused about where you are saving your files, take a moment to understand where they are going. If you are not crystal clear on where your files are stored on your PC for these exercises, stop right now and figure it out.

   Once it starts to unzip, it will look as if it's "doing something." This activity means you're on the right track. You should see eight files for each zipped file, with .shp and .dbf extensions, and this unzipped shapefile, my friend, is solid gold—and ready for mapping.

US CENSUS GEOGRAPHIES CAN BE CONFUSING

The US Census Bureau website allows you to select shapefiles and tabular data for many types of geography (tracts, counties, the entire nation, states, and more). Here is a quick reference to the most widely used geographies:

- Nation: this file is for the US as a whole. If you select this geography and then, for example, population as a data variable, the result will be one number, the population of the entire US.
- State: allows you to select one state, multiple states, or all states.
- County: allows you to select one county, multiple counties, or all counties for the entire US.
- Place: represents city boundaries, plus Census Designated Places.
- Census tract: tracts are the most popular subcounty geography. They are fixed in population between 1,000 and 8,000 people. Census tracts average about 4,000 people, although this amount varies widely.

Adding shapefiles to ArcMap

Open ArcMap

1. **On the Windows Start menu, click the Windows button in the lower-left corner. In the A's section, click ArcGIS to expand it, and from the submenu below it, click ArcMap.** Ignore all those other programs—ArcMap is *the* mapping program, the one that makes the maps.

   The software will take a moment to open, which is normal.

NOTE: *If you get a pop-up window about licensing, you must enter your license into ArcGIS Administrator. The steps for entering your license are in the introduction to this book, under "Authorize Your Trial."*

2. **When a window appears asking you to select a blank map template, click Blank Map (or you can click Cancel, which does the same thing). And then click OK.** If this window does not appear, it is not a problem. You are simply trying to get to a new empty mapping session.

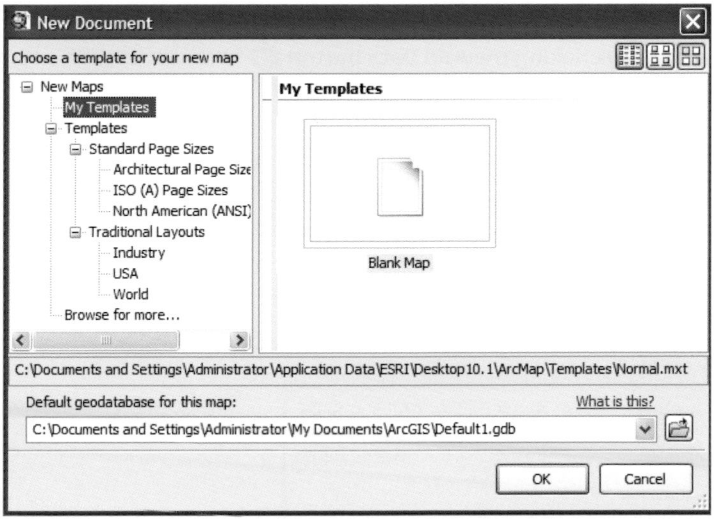

NOTE: *This "Open ArcMap" set of steps is not repeated throughout this book. But you will generally use these steps to start an exercise.*

WHAT FILES MAKE UP A SHAPEFILE?

After unzipping the Census Bureau files, you will have a file with the .shp extension. You will also have a few other files with different file extensions, because a shapefile is made up of multiple files, not just the .shp file.

Mandatory files:

- .shp = the visual image of the map.
- .dbf = database file, in which data is warehoused for the shapefile.
- .shx = index file, which ties the .shp and .dbf files together.

Optional files:

- .prj = projection file, which gives your map its shape, area, direction, and distance (discussed in chapter 3).
- .shp.xml = file contains metadata about your shapefile.
- .sbn and .sbx = spatial index files.
- Other files = a few other file types that can be associated with a shapefile.

Add shapefiles

1. **Add shapefiles to ArcMap by clicking the Add Data button.** You will use this tool constantly, so it might be a good idea to get to know it.

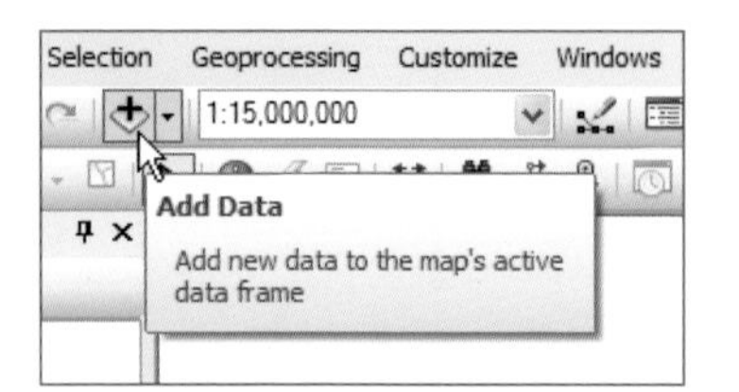

Here is the tricky part: to find your files, you must first do this other thing called "connect to folder"—and there's a button for that.

2. **After clicking the Add Data button, click the Connect to Folder button , and navigate to your save folder (in this example C:\GIS20). Click OK.**

3. **Select tl_2016_01_place.shp, and click Add. Then do the same thing for the county file, tl_2016_us_county.shp.**

   NOTE: *Unless you also downloaded Alabama as your state, you will have slightly different file names. Your state's FIPS code will replace the "01" in the file name.*

4. **If you do not see shapefiles here (they have a .shp extension), you have not properly unzipped the files or they are not in your save folder. Go back and try to unzip and add them again. If you think it's some other issue—it's not.**

FIPS CODES

FIPS code stands for Federal Information Processing Standard code, which provides a unique ID for every parcel of land in the US. States have two-digit codes, and counties have three. So a state plus a county code is a five-digit unique identifier for every county in the US.

HOW DO I FIND MY FILES?

When you first open ArcMap and click Add Data to add shapefiles or other files to ArcMap, the C drive will not be accessible from the navigation list. This inability to get back to the C drive has led to many a frustrated beginner.

The answer is that you must first "connect to folder" to access the C drive. There are two ways to connect to folder:

1. Through the Add Data tool and the Connect to Folder tool .
2. Through ArcCatalog™ and the Connect to Folder tool , or by right-clicking Folder Connections Folder Connections .

Also, you do not have to choose a folder; you can simply navigate to the desktop and connect there. All connected pathways should be evident under the Folder Connections link via the Add Data tool or through ArcCatalog.

Explore essential tools

Because ArcMap provides hundreds of tools, it is essential to identify the most important ones. The tools you will use in nearly every mapping session are featured in this section. Try to become familiar with these tools and what they can do.

Go to Customize > Toolbars, and make sure the following three essential toolbars are turned on: Standard, Tools, and Draw.

The Add Data button

You already know how to add geography files, layer files, or data tables, but let's review: simply click the Add Data button on the Standard toolbar.

1. **Click the Add Data button and get comfortable with using the navigation list to add files, connect to folders, and generally just find your way around. You will notice if you click the little down arrow next to the button, other options appear. Ignore these other options.**

The Zoom In/Zoom Out, Fixed Zoom In/Fixed Zoom Out tools

ArcGIS has four tools for zooming in and out of your map: Zoom In, Zoom Out, Fixed Zoom In, and Fixed Zoom Out.

2. **Click the Zoom In button .**

3. **The tool works better by drawing a square around whatever you want to magnify instead of just clicking the map. For example, activate the tool, and then draw a square with it around your state. Notice how it is easier to control the image by first drawing a square. Now try the other zoom tools, and see what happens.** Do not worry if you mess up your map. The next tool will help fix it.

TIP *You can also easily zoom in and out of your map by using the scroll wheel or trackball on your mouse, if you have a mouse with this functionality.*

The Full Extent tool

The Full Extent tool will resize your map so it fits on your screen. The tool is a great way to center—or recenter—your map.

4. **Click the Full Extent button, and notice how your map is repositioned.**

The Pan tool

The Pan tool looks like a little hand. The tool allows you to reposition the map as if you were moving it on your screen with your hand.

5. **Click the Pan button, and move your map around. To recenter it, click Full Extent again.**

Select Elements (the default pointer)

The default pointer doesn't do anything. That's the beauty of it.

6. **Click the default pointer, and then click your map. Notice that nothing happens. (This tool is technically called Select Elements, which is the visible name when you hover over it. However, do not confuse it with the Select Features tool .)**

TIP ***To "deactivate" any of the other tools, click the default pointer. This button removes the first tool and activates the default pointer.***

The Identify tool

Identify is one of the most useful tools. You can use the Identify tool to click specific geographies and look at the underlying data. Use the Zoom In tool to zoom in closely to a few counties in your state.

7. **Click the Identify tool, and then click a county. Notice that a box with county information appears. Try a few more counties until you are comfortable with this tool. Click the default pointer to get rid of the Identify tool.**

Explore the table of contents

The Table of Contents window, located on the left side of the ArcMap window, is the organizational panel for working with files in ArcMap. Notice two shapefiles listed in the table of contents under Layers in the illustration. Also notice the five buttons at the top of the table of contents: List By Drawing Order, List By Source, List By Visibility, List By Selection, and Options. Hover over any of these buttons in ArcMap, and a description of what that button does displays.

Table Of Contents

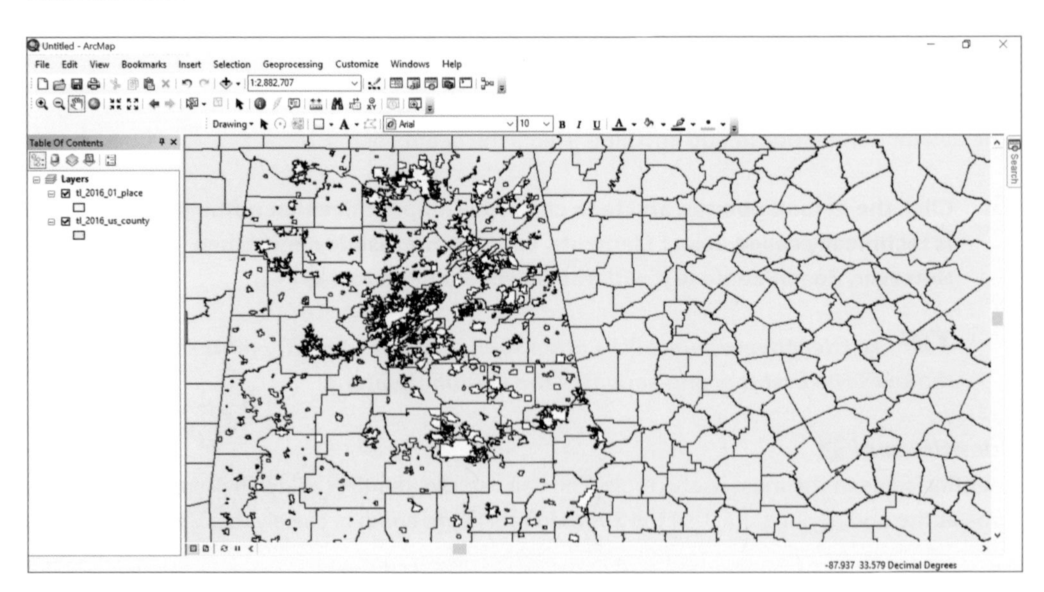

1. **Click the first button, List By Drawing Order** . The default is the second button, List By Source ; however, List By Drawing Order is more useful for most edit sessions.

2. **Use the Zoom In tool to zoom in to your state.**

3. **Display layers by selecting the box next to each layer to turn it on, or by clearing it to turn it off.**

4. **Practice moving the county shapefile and the places shapefile. Move layers up or down by highlighting the layer in the table of contents (clicking the layer once) and dragging it to the desired position.** Notice how your map changes as you reposition the layers. When the place layer is on top, you can see cities in your state. When the county layer is on top, it blocks out the place layer because it is the top layer and has a solid fill color associated with it. Move the place layer into first (top) position.

The layer names are what will be used in the legend. You may want to make them more reader-friendly by renaming them to common names such as Counties and Cities.

TIP ***This functionality will work only while in List By Drawing order mode. If you are unable to move these layers up or down, click the first button, List By Drawing Order, and try again.***

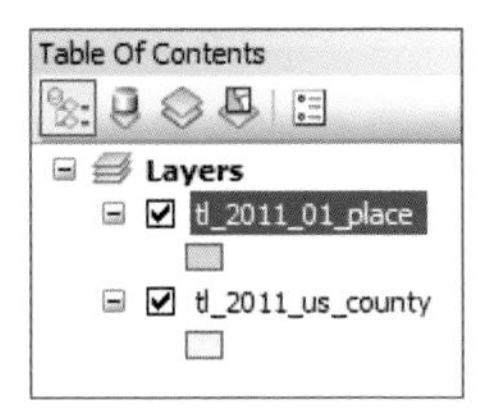

5. **Click the layer name twice to activate the text. Type over the existing layer name.** The layer is not renamed in the underlying data, but only in your map.

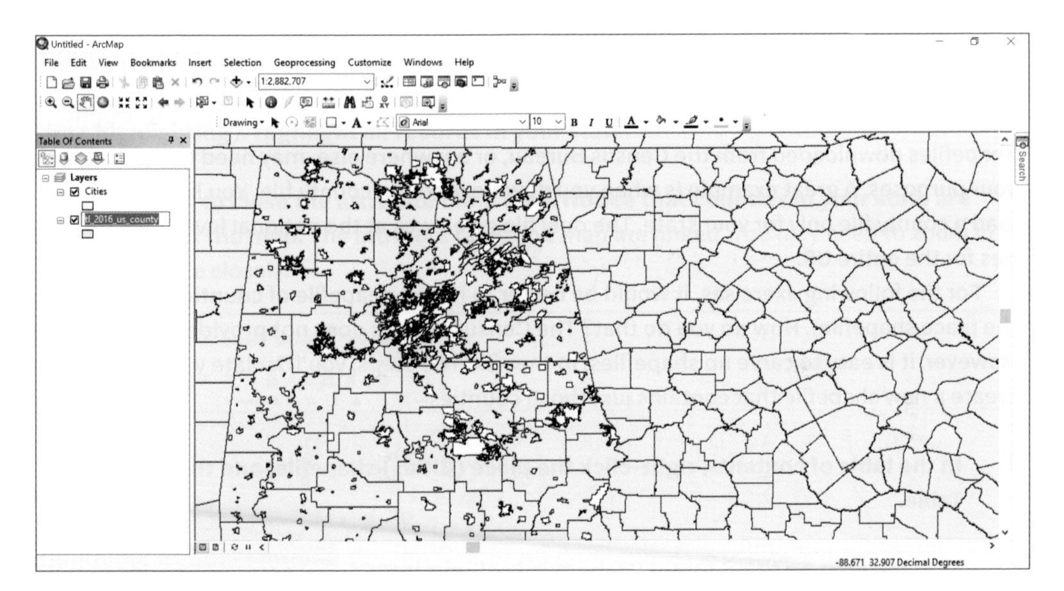

View data

Shapefiles contain two things: the map and the underlying data table. So far, you have looked at the map part of the shapefile and used the Identify tool to look at the underlying data. Another way to view the underlying data is to view the whole data table at once.

1. **In the table of contents, right-click the county layer name, and then click Open Attribute Table. Use the scroll bars (right, bottom) of the attribute table to better understand what is available in the underlying data table.** You should notice a few thousand counties in the

Now you're going to create a new shapefile for just those counties in your state.

7. Right-click the Counties layer name in the table of contents. Click Data, and then click Export Data. Leave all default options, except click the Browse button and navigate to your save folder (under Folder Connections).

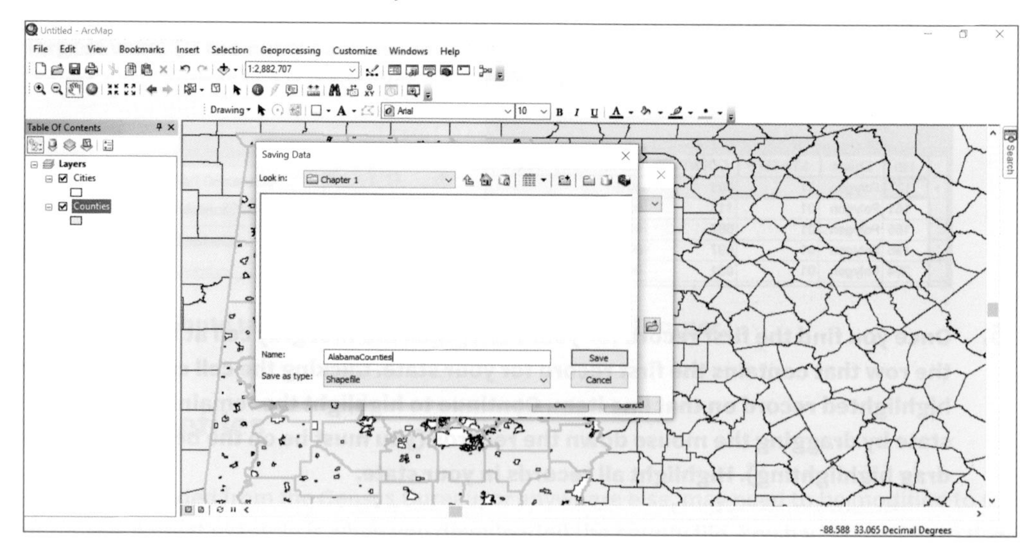

8. Name the file your state's name plus Counties**—for example, "AlabamaCounties." From the "Save as type" list, select Shapefile. Click Save, and then click OK. When prompted with the box that says, "Do you want to add the exported data to the map as a layer?" click Yes. Notice that a new layer appears in the table of contents.**

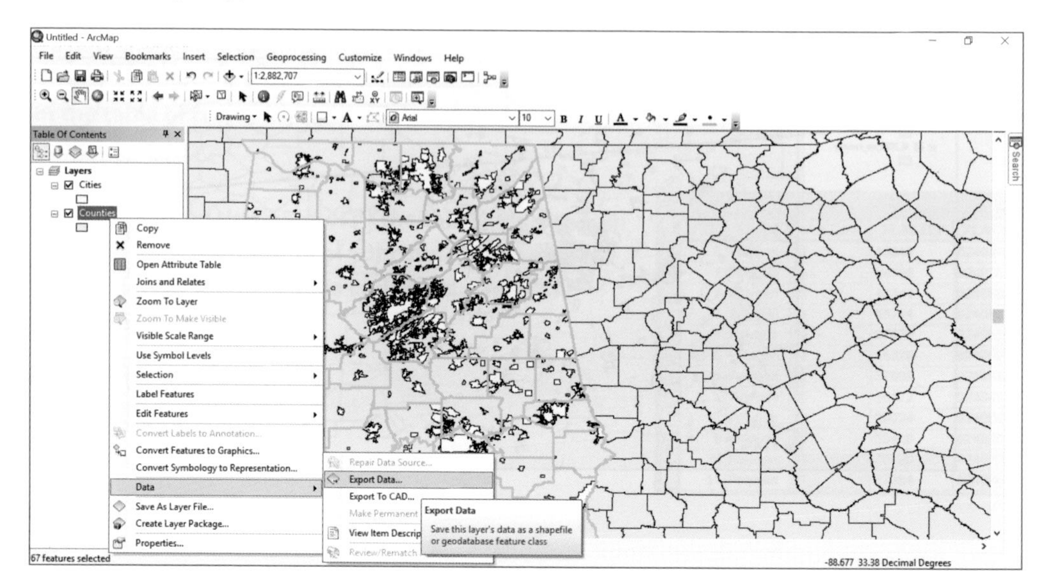

Now you can organize things a bit. The US county layer is no longer needed. So you can remove it.

9. **In the table of contents, right-click the US county layer name (this file is your original county file), and then click Remove. Drag the place file into first position in the table of contents. Notice all the cities in your state, as well as all the counties.**

TIP *Never include spaces in file names.*

Save the project

1. **On the File menu, click Save As.**
2. **Navigate to your save folder.**
3. **Type a name for this project. Call it** mystate.mxd**.**
4. **Click Save. Notice in the upper-left corner, your workspace has been given this new name.**

SAVING A PROJECT
Saving, although technically easy, is more complicated than it appears. When you "save" a project, ArcMap saves the project as an .mxd file or ArcMap document. The .mxd file functions as a pointer file to all the files that make up the project. If you sent the .mxd file to someone, they would not be able to open it without all the files that make up the project.

5. **Click the x in the upper-right corner to close ArcMap.**

CHAPTER 2

KEY CONCEPTS

working with multiple layers
changing map colors
creating labels
creating layouts
creating legends
creating titles
using scale bars
using north arrows

Creating basic maps and layouts

Reference maps are basic, traditional maps like those you see in atlases. Their purpose is to illustrate geographic boundaries of a given area, such as cities or counties. These types of maps are the cornerstone of cartography. Layouts contain titles, legends, north arrows, scale bars, and many other graphic features. It is important that you learn to create layouts that help your reader quickly understand your map.

In this exercise, you will make your first map.

You will need to download all chapter data files from esri.com/GIS20-3.

Make a layer semitransparent

1. **In the table of contents, right-click the Cities layer, and then click Properties.**

2. **Click the Display tab.**

3. **On the right of Transparent, type** 50**. This designation will make the Cities layer 50 percent transparent, allowing you to see through it. Click OK. Notice how the layer is lighter.** You did not need to make this layer semitransparent; however, it makes the map slightly easier to read and teaches a new skill.

Layer Properties
General | Source | Selection | Display | Symbology
Scale symbols when a reference scale is set
Transparent: 50 %
Display Expression

Reorder layers

1. **In the table of contents, drag the county layer into first position. If you are unable to do this, ensure that you select List By Drawing Order (the first button) in the table of contents.** This step makes a subtle change. The counties are in the top layer, with light-gray lines, and the semitransparent cities are displayed underneath.

TIP *For a reference map, it is difficult to illustrate more than four layers on one map without the map becoming too cluttered.*

Turn on labels

Labeling is essential to creating a quality reference map. So now you can label cities.

1. **In the table of contents, right-click the layer you want to label (Cities), and then click Properties.**

2. **Click the Labels tab. Select the "Label features in this layer" box, which is a tiny little check box in the upper-left corner.** The appropriate field for labeling is already selected by default (Name).

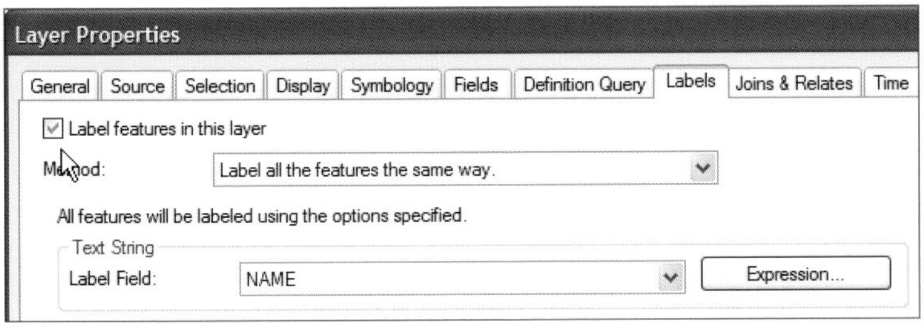

3. **Click OK, and have a look at the labels.** What a mess! Now you can fix these labels, and make the map easier to read.

Fix labels

You can do many things to improve messy labels:

- Remove duplicate labels.
- Assign a buffer to labels, which places only some of the labels.
- Create a halo (a white outline) around labels.

To remove duplicate labels, complete the following steps:

1. **In the table of contents, right-click Cities, and then click Properties.**

2. **Click the Labels tab.**

3. **In the lower-left corner, click Placement Properties.**

4. **At the top, click the Placement tab.**

5. **Select the Remove Duplicate Labels option.**

To assign buffers to labels, complete the following steps:

6. **At the top, click the Conflict Detection tab.**

3. **You cannot leave the map looking like this, so switch back to data view (View > Data View), and use the Full Extent tool to resize the map to something more appropriate.**

4. **Return to layout view (View > Layout View).**

Create a layout

Depending on whether the geography is horizontal (Nebraska) or vertical (California), you'll change the map's orientation to either portrait or landscape to accommodate the geography shape.

1. **To change the orientation from the default (portrait) to landscape, right-click in the white space anywhere outside of the print area (or what looks like the edge of a piece of paper) in the layout.**

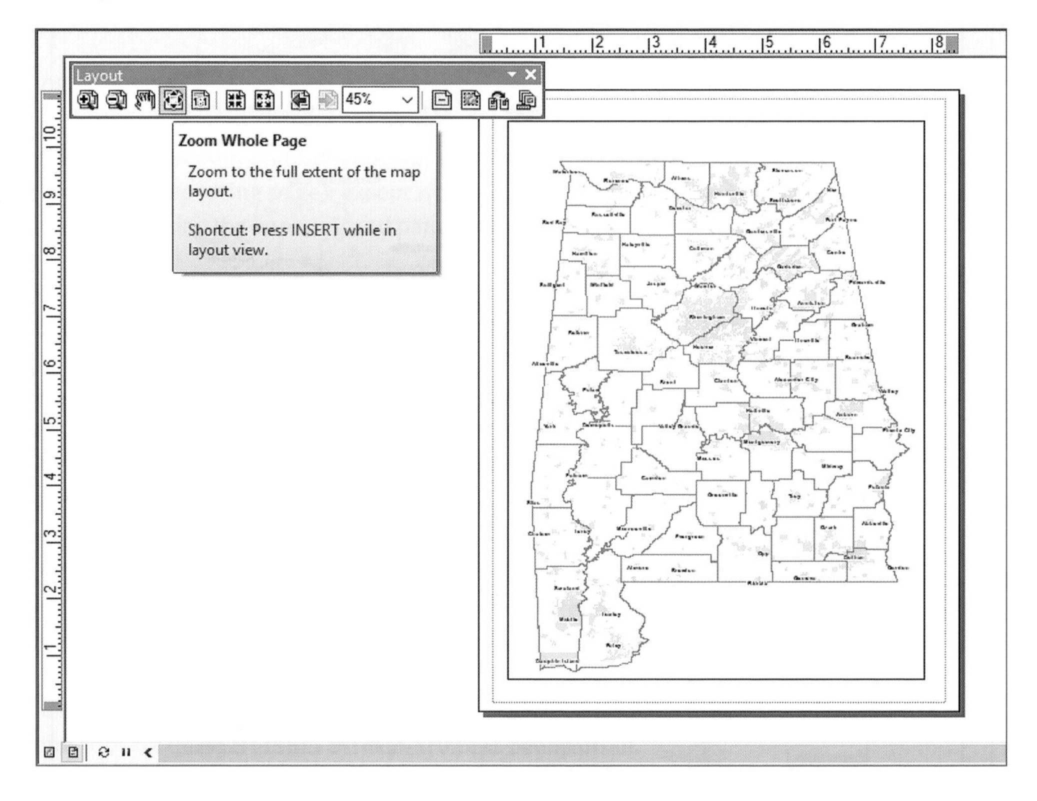

2. **Click Page and Print Setup.**

3. **On the right of Orientation, select the Landscape option.**

4. **Select the "Scale Map Elements proportionally to changes in Page Size" check box (lower right) to center the map in the page layout. Click OK.**

5. **Resize the image to fit the space by clicking the map once to select it, and then dragging the dots in each of the four corners.** Do not worry about skewing the map when you drag it—ArcMap keeps the correct proportions of your geography.

6. **Click one of the little dots on the corner until a double-headed arrow appears. Drag the image to the desired position, and repeat this action with each corner dot until the map is as large as you want it.** Sometimes, it is hard to get positioned correctly. You can still use the Pan and Magnifier tools to center the image and make it larger. The key thing to know is that the rectangular box around the map is called a *map frame* and should be inside the outer edge of the largest rectangle, which represents an 8½ x 11 inch piece of paper. The map frame is the print area.

THE PRINT AREA

In layout view, the print area is defined by the default rectangle that looks like a piece of paper. This rectangle represents an 8½ x 11 inch piece of paper. Stay within the boundaries of this rectangle. Anything outside the boundaries will not print.

A landscape page format allows for a larger map image for most geography; however, you'll decide whether your state is best suited for landscape or portrait.

In this example, Alabama looks better as a portrait, so I'm going to switch it back. Feel free to do the same now if you prefer your map to be portrait.

7. **Optional: to switch back to portrait, resize your map again by dragging the corner dots and using the Full Extent tool to resize it.**

TOOLS TOOLBAR STILL WORKS

Notice that all the tools on the Tools toolbar are still available and active, even in layout view. The Full Extent tool is particularly helpful when working in layout view.

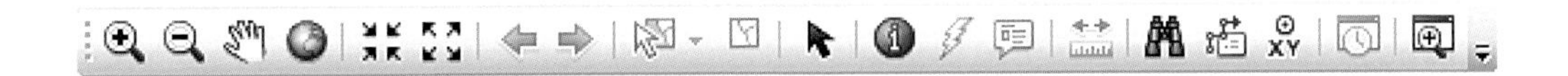

TIP *The layout has a new toolbar. One of the best tools on it is Zoom Whole Page . It recenters everything in case things start to get off base.*

ORGANIZING THE LEGEND

When you work with multiple layers, the legend should be organized in order of scale (from largest to smallest—for example, state, county, and city). Another good idea is to put land masses (physical land such as state, county, city) at the top of the legend and other context layers, such as streets and water, at the bottom.

Insert a scale bar

Traditional cartography instructs that scale bars must accompany every map; however, scale bars are frequently left off thematic maps (chapter 6) because they serve no real purpose on a thematic map. Scale bars are required on reference maps because distance is an important element.

1. **On the Insert menu, click Scale Bar.**

2. **Select the first scale bar by clicking it once, and then click OK. If you are using an older version of ArcGIS Desktop, you must modify scale bar properties and change division units from decimal degrees to miles.** The scale bar is dropped right in the middle of the map.

3. **Move the scale bar to the lower-right corner.** The easiest way to change the mileage of the scale bar is to drag the right-side dot of the scale bar box and widen the scale bar. If the map size changes, the scale bar automatically adjusts. It is not a fixed element.

4. **Widen the scale bar to show 100 miles.**

Insert a north arrow

With north arrows, the simpler the better. Avoid overly ornate arrows.

1. **On the Insert menu, click North Arrow.** The font is difficult to change on ArcMap north arrows.

2. **Select a north arrow with a font similar to fonts used in the map, ideally a sans serif font.** Esri North 6 is a good option.

3. **Move the north arrow into the legend box, and place it in the upper-right corner. Make the arrow smaller so it fits in the box.**

Insert source using text

Text is a versatile tool and can be used to provide a source. Providing data sources is a must, but you may also consider providing sources for shapefiles as well as the map itself. There are many ways to cite sources. One recommended way is the American Psychological Association (APA) method. Many fields use this method, not just psychology.

1. **To insert text, on the Insert menu, click Text.** The text box is dropped in the center of the map, which makes it difficult to see.

2. **Click anywhere outside the text box to deactivate it, and then hover over the text box. A four-pronged arrow will appear. Drag the text box to the lower-right corner so it is easier to see and work with.**

3. **Right-click the text box, and then click Properties (double-clicking also works).** You can have a map source (your agency) as well as a data/geography source (the US Census Bureau).

4. **Type the following information:**
   - Map Source: Your organization, June 2017 **(or the correct month and year, of course).**
   - Shapefile Source: US Census Bureau. (2016, January 1). 2016TIGER/Line Shapefiles for: Alabama. Retrieved June 28, 2017, from US Census Bureau: http://www.census.gov/cgi-bin/geo/shapefiles/index.php.

   Notice the font is Arial 10. Arial is a fine sans serif font, but make the font larger.

5. **Click Change Symbol, and change the font size from 10 to 12. Click OK.**

   TIP ***Sources, as well as other map text, should generally not be smaller than 10-point font.***

6. **Click the Left Alignment button, to align text left. You must do this step last—otherwise, it will not work. Click OK.** Notice the citation is too long and runs off the page.

7. **Right-click the text box, and then click Properties.**

8. **Insert the cursor where you want to break the line, and then press Enter. You may need to insert four or five line breaks so the text doesn't run too far to the right.** The line breaks help stack the lines on top of each other. You can click Apply each time to see how it looks.

9. **Once the text looks reasonable, click OK.**

MAP FLOW

The philosophy behind map flow is based on the principles of photography. The eye should be drawn to certain anchor points and move across the page. In the example in the figure, the eye moves in a *Z* pattern. Avoid clustering all elements on one side of the map, which creates an unbalanced composition. White space should be fairly equal on all sides of the map.

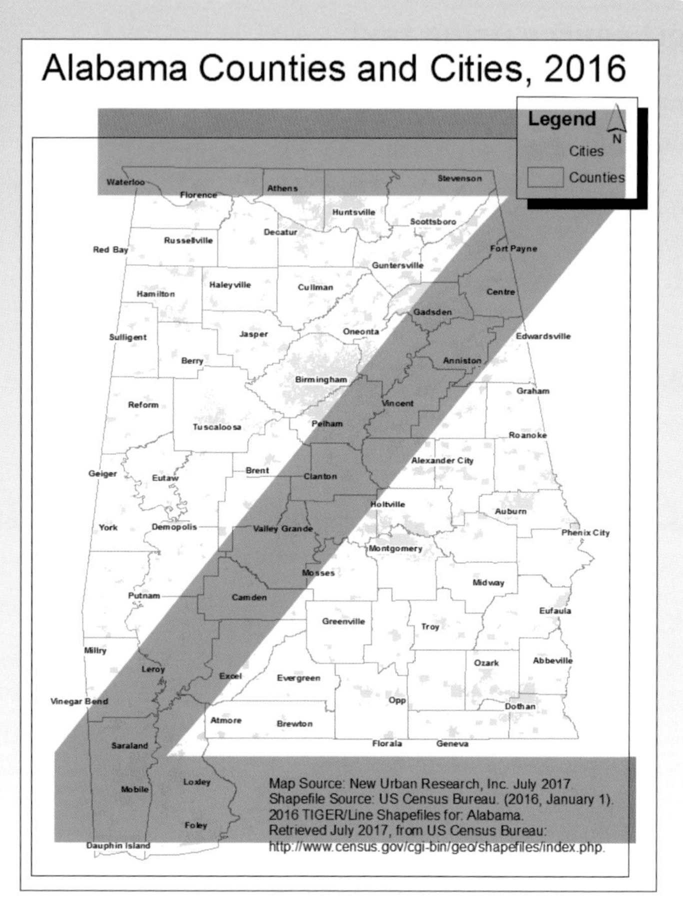

Save the project

1. **On the File menu, click Save As.**
2. **Navigate to your save folder.**
3. **Name the map for your state, such as** Alabama.mxd**. Click Save. Close ArcMap, or to proceed to the next exercise in the next chapter, click File > New > Blank Map. And click OK.**
4. **Go get a snack.**

CHAPTER 3

KEY CONCEPTS

changing projections
understanding UTM
understanding state plane and custom projections
recognizing and solving common projection problems

Projecting shapefiles

Projections give shapefiles the correct shape, area, direction, and distance. Defining the projection for shapefiles will ensure that the geography is reflected properly, that distance is recorded accurately, and that all layers are visible. There are easier things to understand than map projections. This chapter seeks to provide common and practical solutions for typical projection issues, not present a dissertation on the subject.

You will need to download all chapter data files from esri.com/GIS20-3.

I'm not going to lie—learning how to project shapefiles isn't the most glamorous part of GIS, but it is hugely important. In this exercise, you will not only learn how to solve basic projection issues but project shapefiles yourself.

DISTANCE ACCURACY AND AVOIDING AWKWARD CONVERSATIONS

A good starting point for a discussion on accuracy is to ask how accurate your map needs to be. For basic mapping, especially for demographics and other thematic maps, your map likely does not need to be close to reality, or within a hair's breadth of the true distance. However, being 50 miles off course probably won't work either. If distance accuracy is even a little important for your project, consider having a discussion with your team to agree on what perspective is appropriate, which could save many awkward discussions later.

The second thing you should do is make sure that all shapefiles for one analysis are the same projection so they will all line up with each other. For example, you may have a layer of bus stops and a layer of streets. You publish this map to the web. Then you look at it closely and realize the stops are not at the correct locations. They are slightly off or, in the worst case, really off. In the meantime, thousands of unsuspecting bus riders have looked at your map, and they all went to the wrong corners.

Understanding the basics

Coordinate systems can be grouped into geographic coordinate systems (GCS) and projected coordinate systems (PCS). The key distinction is whether they use a 3D object (think *globe*) or a 2D, flat surface (think *map*) to represent the surface of the earth.

Many coordinate systems are used; however, latitude-longitude is the oldest (developed before the birth of Christ) and uses global lines running north–south (latitude) and east–west (longitude) to assign a set of numbers to every location on Earth. Latitude-longitude is based on a 3D spherical surface and is the foundation of many coordinate systems.

The problem with projections comes when you must represent a 3D object on a 2D surface. This situation is where a PCS comes in handy. Universal transverse Mercator (UTM) is a PCS developed in the 1940s by the US Army Corps of Engineers. UTM uses a grid system to assign numbers to grid cells worldwide. It divides the surface of the earth into 60 zones and relies on a flat, 2D model.

A *datum* is a set of constant, known points used to model the earth's surface in a particular area, such as North America. Many datums exist because many people all over the world had the same idea to collect and analyze points to create a representative model of the earth.

In North America, the most commonly used datum is NAD 1983, which is based on 250,000 points across North America. In 2007, NAD 1983 was updated, and the resulting datum was updated to NAD 83 (NSRS2007). HARN (High Accuracy Reference Network) is a statewide or regional upgrade in accuracy of NAD 1983. Another widely used datum is the World Geodetic System 1984 (WGS84), which the US Department of Defense uses.

Census shapefiles are already in NAD 1983; however, North America is a big place, and many GIS projects are focused on a much smaller area, such as county level or even neighborhood level. Therefore, applying an appropriate *projection* provides a better visual representation of local geography and makes distance more accurate.

It would be nice if there was one quick, easy way to do projections. But there isn't. You must apply various projections depending on the location's geographic scale and where on the earth you are mapping.

A GUIDE TO MAP PROJECTIONS
If instead of just one chapter on projections, you want to read a whole book about the subject, you are in luck. *Lining Up Data in ArcGIS*, second edition, by Margaret Maher (Esri Press, 2013) discusses all things related to map projections, including creating custom map projections.

Projecting shapefiles when the geographic coordinate system is known

In chapter 1, you downloaded shapefiles from the US Census Bureau. They come in NAD 1983. What if you didn't already know this? Or what if you get shapefiles from another source, and don't know anything about their projection? Following are the steps to identify a shapefile's coordinate system, datum, and projection.

Get datum information

1. **Open ArcMap. (If you don't know how to do that, see chapter 1, "Open ArcMap.") It's important to start a fresh map. If you did not just open ArcMap, perhaps because you were working on chapter 2, go to New Maps > Blank Map to get a blank map.**

2. **Click Add Data, navigate to your save folder, and select the file created in chapter 1 named AlabamaCounties (yours will likely have a different name), which contains all counties in your state.**

NOTE: *Do not select the counties file containing all the counties for the US. That file requires a different projection.*

Because you must apply a state-level projection, it is important to figure out which projection to use. The projection varies depending on state.

Look up state plane projections for your state

1. **Open an internet browser, and navigate to the State Plane Coordinate System Designations website created by Rick King, http://stateplane.ret3.net. This site provides a quick and easy way to determine which projection is best.**

2. **Scroll down to your state, find your county in the list, and jot down the FIPSZone.** The FIPS (Federal Information Processing Standard) zone denotes the state plane projection. Note that UTM is also a valid projection. For ease of explanation, the state plane is used. If you know that your state or county uses UTM as the preferred projection, use that.

   Alabama is covered by FIPS zones 0101 and 0102, which relate to the east and west sides of the state, respectively. What this means is that for state plane projection, the state is split into two zones. California, on the other hand, is split into six zones, and Connecticut has only one zone.

TIP ***In this exercise for ease of explanation, we're reprojecting to state plane rather than UTM. Using UTM would never be considered wrong, but your state, county, city, or office GIS person may prefer you to use some other projection, and that's okay.***

STATE PLANE VERSUS UTM

Curious what the difference is in state plane versus UTM?

These two figures illustrate the difference in UTM versus state plane. Colorado falls mostly within UTM 13 and is evenly distributed between state plane zones 0501, 0502, and 0503.

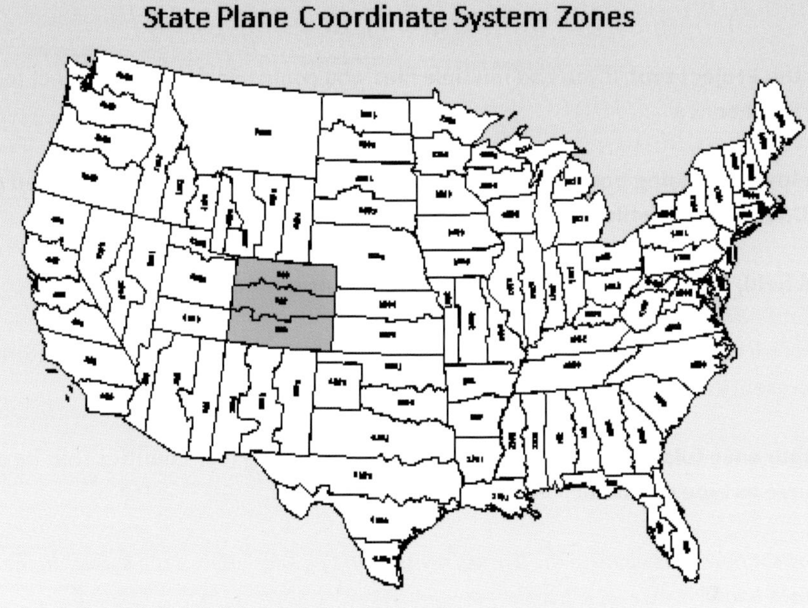

Colorado is evenly distributed among three state plane zones.

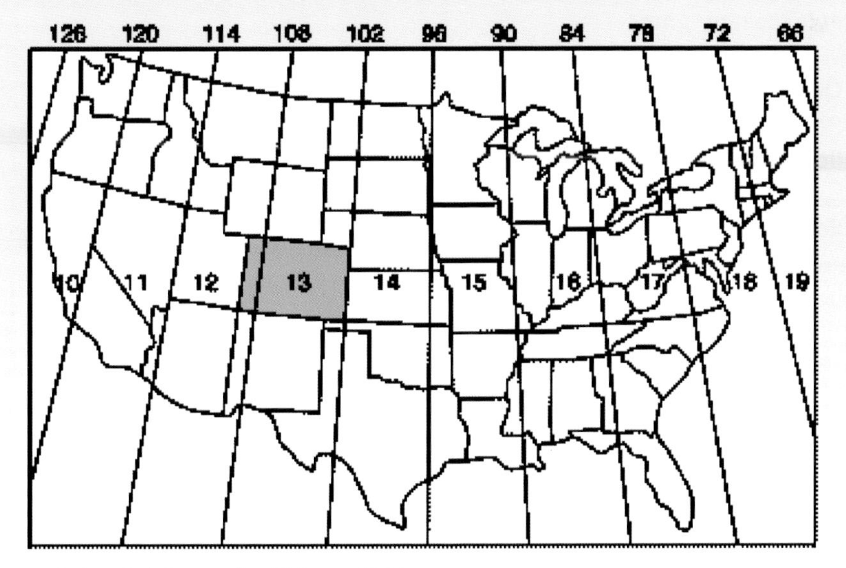

Colorado falls mostly within one UTM.

Project your file

1. **Click the ArcToolbox button to open ArcToolbox.**

2. **Expand the Data Management Tools toolbox and then the Projections and Transformations toolset.**

3. **Double-click the Project tool.** If you had multiple files, you could use the Batch Project tool to project many files at once.

4. **In the first field of the dialog box, click the Input Datasets or Feature Class arrow, and select the AlabamaCounties shapefile.**

5. **In the second field, Input Coordinate System will fill in automatically.**

   The next step is tricky so make sure you're paying attention. The Output Dataset or Features Class field represents where to save what will be a newly projected shapefile.

6. **Navigate to your save folder, and name the shapefile** countiesprj **(for counties that have been projected). Save as type should be Feature classes. Click Save.**

   NOTE: *If it is unclear how to navigate to the C drive, revisit "Finding Files—Here's the Secret" in the introduction of the book.*

   Now we get to the nitty-gritty.

7. **Click the Output Coordinate System button .**

8. **Double-click the Projected Coordinate Systems folder.**

9. **Double-click the State Plane folder.**

10. **Double-click NAD 1983 (US Feet).**

11. **Select the correct state plane (zone) for your state (the one you looked up a second ago). Click OK to close each window. Wait for the green pop-up check box to appear in the lower-right corner.** When you see the check box, you know the projection worked. Running the reprojection could take a full minute.

12. **Close ArcToolbox since it's probably right in the middle of your screen.**

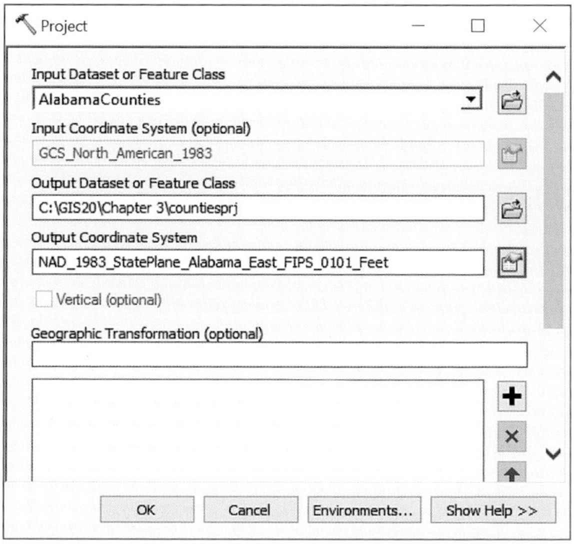

Add file to ArcMap

Here is some next-level cool stuff to know. After the shapefile reprojects and you get the green check mark, you must open a new blank map to add the newly reprojected shapefile. I have seen students skip this step of opening the new blank map and simply add the reprojected shapefile to the edit session they are in. When this happens, the reprojected shapefile does not look any different, because it's conforming to the projection of the edit session you are already in. Now, you will complete a little-known step—*opening a new ArcMap edit session before you can see any changes*. You probably already know how to open a blank map by now, but if not, you can do these steps.

1. **In the upper-left corner, go to File > New, and then select Blank Map (if the option is given). Click OK. When prompted to "Save Changes to Untitled?" click No. You should see a blank, empty workspace.**

2. **Use the Add Data button to add the new projected file countiesprj.shp to ArcMap.** Notice how the outline of the state looks slightly different. In some cases, depending on the shape of the state, it might look very different. You now have a correctly projected shapefile for your little corner of the world. Congratulations!

3. **Close ArcMap, or to proceed to the next exercise, click File> New > Blank Map, and click OK. There's no need to save.**

CHAPTER 4

KEY CONCEPTS

preparing Microsoft® Excel files for ArcMap
working with census data

Preparing data for ArcMap

Data is the fuel that maps use. Most aspiring mappers want to get their data into a map. To do that, it's important to learn the tips and tricks that make it possible to pull Excel spreadsheets into ArcMap. Another awesome thing to learn is how to get some of the best available data in the US—census data—and then prepare this data for mapping (ahem, job security).

[*You will need to download all chapter data files from esri.com/GIS20-3.*

The Census Bureau does a census of the US every 10 years. This decennial count used to be the ultimate in terms of data. However, several years ago, the Census Bureau created another survey called the American Community Survey (ACS), which is conducted annually instead of decennially. Hundreds of data tables are available that provide all sorts of information about the US population.

Although you will use a census data table to illustrate data preparation for ArcMap, this exercise is directly applicable to any Excel spreadsheet you want to map. Many tips, tricks, and techniques essential to working with Excel spreadsheets in ArcMap are included.

In this exercise, you will download the percentage of senior citizens for counties in your state. This data happens to be age related, but the whole process works the same for population, race and ethnicity, unemployment, income, housing, and all sorts of stuff. Learning how to get ACS data and map it is a solid-gold party trick. It also will make you look like a genius to your boss, and could possibly help solve some real-life problems.

Downloading data from the US Census

Navigate to American Factfinder

1. **Go to https://www.census.gov.**

2. **Click the Data tab (on the upper toolbar), and then click the Data Main button. Then click the American FactFinder link.** Think of American FactFinder as the secret vault warehousing all sorts of current, free, trustworthy data.

3. **Click Advanced Search, and then click Show Me All.**

Select geography

Your ultimate goal is to map data you are downloading now. Eventually, this data will be joined to the countiesprj shapefile, which includes all the counties in your state. First, you'll select the geographic type, which, again, is county level.

It's great to know that when you download census data, you will always begin with the geographic selection first. The reason is, the geographic level you select determines what data tables are available to you. The lower level the geography, the more census surveys it takes to get the data accurate, and the more surveys required, the longer it takes to get the data at that level. In other words, national-level data is released first for the whole country, and then other releases come out, eventually getting down to the census tract level, which is the lowest level of geography for which ACS data is available. This process is ironclad—geography, then data.

1. **In the left column, click Geographies.**

2. **In the "Select a geographic type" list, click County - 50.**

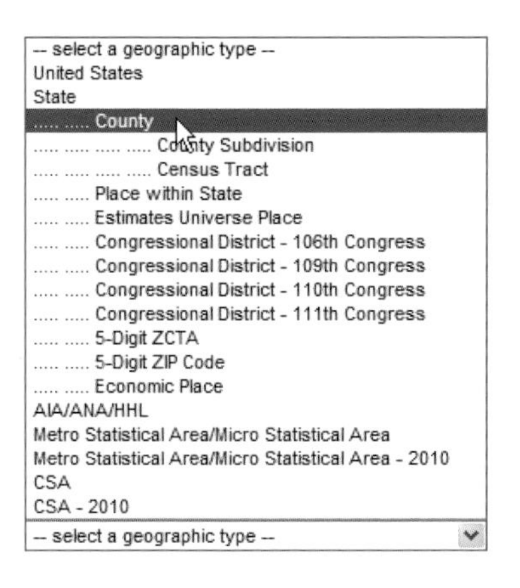

3. **In the "Select a state" list, click your state.**

4. **In the last list, click "All Counties within <your state>," and then click Add to Your Selections.**

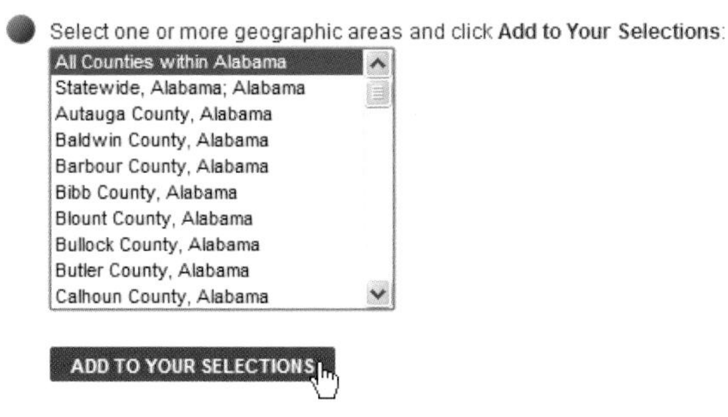

5. **Notice in the upper-left corner under Your Selections, this geography has been recorded.**

6. **Close the dialog box by clicking Close in the upper-right corner—we're not talking about the whole browser, just the pop-up window.**

Search data

You must now choose the data tables to map. In this exercise, you'll map senior citizens (persons 65 years old and older) by county for your state.

1. **In the Topic or Table Name search box at the top of the page, type** age **and click Go.**

Several results are returned, but the ones the Census Bureau thinks are the best match (and are most frequently used) are indicated with a star and are at the top of the list. You may have noted that 2016 data is available. But here's the catch—not at the county level (yet). Depending on when you work on this exercise, the data could be updated.

Either way, you'll know the currency of the data you're working with because the Census Bureau courteously always puts the most current data at the top of the list. For the table Age and Sex (at the county level), it's currently only available for 2015. So because the current data is always at the top, you don't have to wonder if you are accessing the most current data.

2. Select the check box for the row beginning with S0101 Age and Sex, and make triple sure it's the "2015 ACS 5-year estimates" indicated on the right side of the page in the same row.

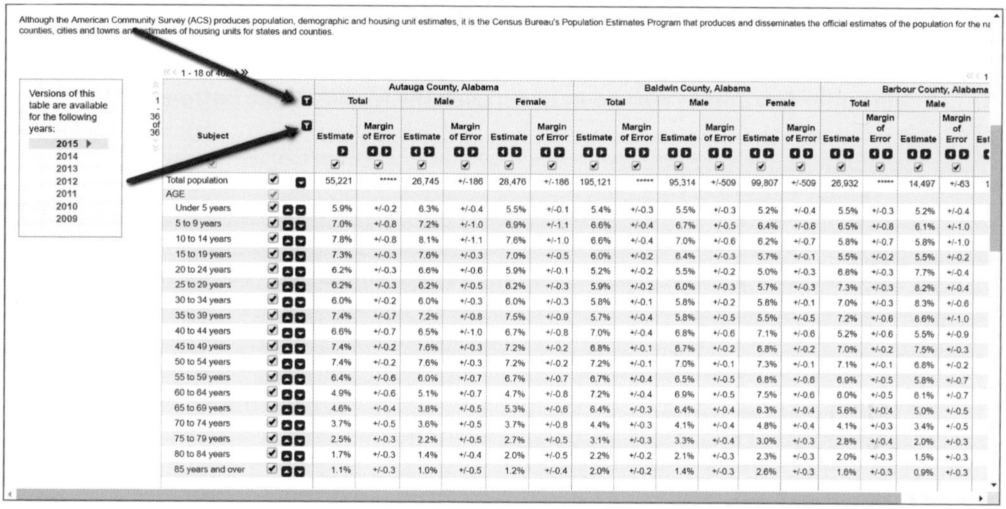

Although the American Community Survey (ACS) produces population, demographic and housing unit estimates, it is the Census Bureau's Population Estimates Program that produces and disseminates the official estimates of the population for the na counties, cities and towns an estimates of housing units for states and counties.

Versions of this table are available for the following years:
2015
2014
2013
2012
2011
2010
2009

1 - 18 of 40

Subject	Autauga County, Alabama						Baldwin County, Alabama						Barbour County, Alabama			
	Total		Male		Female		Total		Male		Female		Total		Male	
	Estimate	Margin of Error	Estimate	Margin of Error	Estimate	Margin of Error	Estimate	Margin of Error	Estimate	Margin of Error	Estimate	Margin of Error	Estimate	Margin of Error	Estimate	Margin of Error
Total population	55,221	*****	26,745	+/-186	28,476	+/-186	195,121	*****	95,314	+/-509	99,807	+/-509	26,932	*****	14,497	+/-63
AGE																
Under 5 years	5.9%	+/-0.2	6.3%	+/-0.4	5.5%	+/-0.1	5.4%	+/-0.3	5.5%	+/-0.3	5.2%	+/-0.4	5.5%	+/-0.3	5.2%	+/-0.4
5 to 9 years	7.0%	+/-0.8	7.2%	+/-1.0	6.9%	+/-1.1	6.6%	+/-0.4	6.7%	+/-0.5	6.4%	+/-0.6	6.5%	+/-0.8	6.1%	+/-1.0
10 to 14 years	7.8%	+/-0.8	8.1%	+/-1.1	7.6%	+/-1.0	6.6%	+/-0.4	7.0%	+/-0.6	6.2%	+/-0.7	5.8%	+/-0.7	5.8%	+/-1.0
15 to 19 years	7.3%	+/-0.3	7.6%	+/-0.3	7.0%	+/-0.5	6.0%	+/-0.2	6.4%	+/-0.3	5.7%	+/-0.1	5.5%	+/-0.2	5.5%	+/-0.2
20 to 24 years	6.2%	+/-0.3	6.6%	+/-0.6	5.9%	+/-0.1	5.2%	+/-0.2	5.5%	+/-0.2	5.0%	+/-0.3	6.8%	+/-0.3	7.7%	+/-0.4
25 to 29 years	6.2%	+/-0.3	6.2%	+/-0.5	6.2%	+/-0.3	5.9%	+/-0.2	6.0%	+/-0.3	5.7%	+/-0.3	7.3%	+/-0.3	8.2%	+/-0.4
30 to 34 years	6.0%	+/-0.2	6.0%	+/-0.3	6.0%	+/-0.3	5.8%	+/-0.1	5.8%	+/-0.2	5.8%	+/-0.1	7.0%	+/-0.3	8.3%	+/-0.6
35 to 39 years	7.4%	+/-0.7	7.2%	+/-0.8	7.5%	+/-0.9	5.7%	+/-0.4	5.8%	+/-0.5	5.5%	+/-0.5	7.2%	+/-0.6	8.6%	+/-1.0
40 to 44 years	6.6%	+/-0.7	6.5%	+/-1.0	6.7%	+/-0.8	7.0%	+/-0.4	6.8%	+/-0.6	7.1%	+/-0.6	5.2%	+/-0.6	5.5%	+/-0.9
45 to 49 years	7.4%	+/-0.2	7.6%	+/-0.3	7.2%	+/-0.2	6.8%	+/-0.1	6.7%	+/-0.2	6.8%	+/-0.2	7.0%	+/-0.2	7.5%	+/-0.3
50 to 54 years	7.4%	+/-0.2	7.6%	+/-0.3	7.2%	+/-0.2	7.2%	+/-0.1	7.0%	+/-0.1	7.3%	+/-0.1	7.1%	+/-0.1	6.8%	+/-0.2
55 to 59 years	6.4%	+/-0.6	6.0%	+/-0.7	6.7%	+/-0.7	6.7%	+/-0.4	6.5%	+/-0.5	6.8%	+/-0.6	6.9%	+/-0.5	5.8%	+/-0.7
60 to 64 years	4.9%	+/-0.6	5.1%	+/-0.7	4.7%	+/-0.8	7.2%	+/-0.4	6.9%	+/-0.5	7.5%	+/-0.6	6.0%	+/-0.5	6.1%	+/-0.7
65 to 69 years	4.6%	+/-0.4	3.8%	+/-0.5	5.3%	+/-0.6	6.4%	+/-0.3	6.4%	+/-0.4	6.3%	+/-0.4	5.6%	+/-0.4	5.0%	+/-0.5
70 to 74 years	3.7%	+/-0.5	3.6%	+/-0.5	3.7%	+/-0.8	4.4%	+/-0.3	4.1%	+/-0.4	4.8%	+/-0.4	4.1%	+/-0.3	3.4%	+/-0.5
75 to 79 years	2.5%	+/-0.3	2.2%	+/-0.5	2.7%	+/-0.5	3.1%	+/-0.3	3.3%	+/-0.4	3.0%	+/-0.3	2.8%	+/-0.4	2.0%	+/-0.3
80 to 84 years	1.7%	+/-0.3	1.4%	+/-0.4	2.0%	+/-0.5	2.2%	+/-0.2	2.1%	+/-0.3	2.3%	+/-0.3	2.0%	+/-0.3	1.5%	+/-0.3
85 years and over	1.1%	+/-0.3	1.0%	+/-0.5	1.2%	+/-0.4	2.0%	+/-0.2	1.4%	+/-0.3	2.6%	+/-0.3	1.6%	+/-0.3	0.9%	+/-0.3

3. At the top (or bottom), click View.

TIP *For the most current available data, the ACS provides one- and five-year estimates for many data tables. The one-year estimate is just that, a tally of the results for the previous year's ACS. It has a much smaller sample set and is the least accurate. The five-year estimate, a rolling average of the past five years, is the most accurate and, without exception, the correct dataset to use. A good way to phrase the use of this data is "and now, folks, as you can see in this map, which uses the most current data available . . ."*

Download data

Notice all the data. What you're after though is the "65 years and over" variable toward the bottom. Instead of exporting all this data, you're going to isolate and export only the variable you want.

1. In the upper-left corner next to Actions, click the Modify Table button.

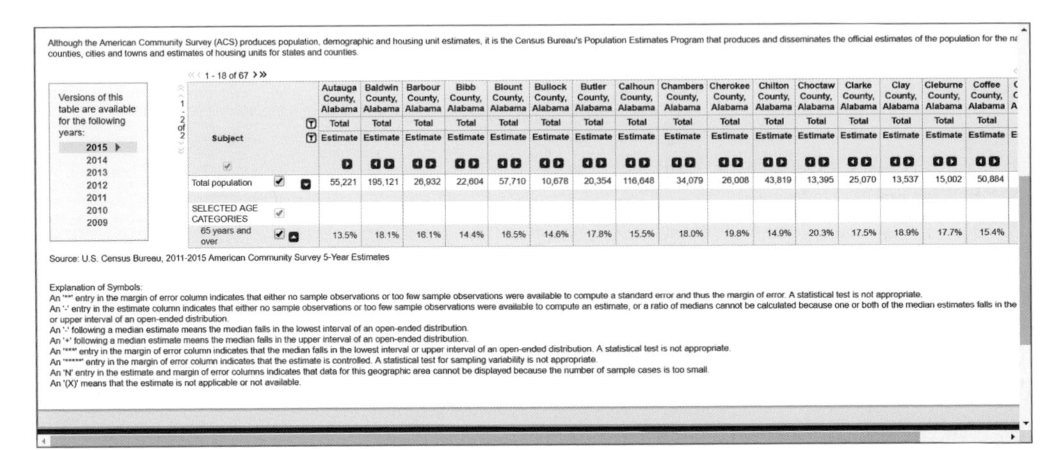

Things get a little tricky now. The ACS seems obsessed with making sure you know how accurate it is. That's what all this Margin of Error business is about. It lets you know what the upper and lower confidence interval ranges are. That's good. But you don't really need it. Also, this data is broken down by males and females, and you also do not need that level of granularity. You ultimately want just a total percentage of seniors by county. So to drastically simplify this table, you can filter.

2. Click the little filter buttons in the Subject square. On the first pop-up filter, select the check box for HC01 Total and click OK, and for the second filter, select the check box for EST Estimate and click OK.

Next, you want to clear all the check boxes on the left for all the age groups you do not want. Unfortunately, there's no easy way to do this. You must simply clear them, one by one.

3. Leave the first one, Total Population, checked, just because. Then the only other one to leave checked is "65 years and over" about halfway down the page. Clear all the other check boxes. In the end, the image should look like the figure in step 1.

NOTE: *If things get messed up, you can start over by clicking the Reset Table button. Or if you accidentally click and turn something off that you want to turn back on, click the Show Hidden Rows/Columns button to view unchecked items, and then simply reselect them.*

4. **Click the Download button on top to download the file. When prompted, click Use, leaving all options checked, and then click OK. The file will begin to download so click Download again to save it. And save it to your save folder.**

5. **Navigate to your save folder, and unzip the file. (If you don't know how to do this, see chapter 1, "Download county, city, and state shapefiles.")**

Import into Excel

After the file is unzipped, there are four files. The file you are interested in is ACS_15_5YR_S0101_with_ann.csv.

Normally, you might simply double-click the file to open it in Excel. There are some strange things at work here, though, regarding the FIPS code. *Do not simply double-click the file to open it.* You must import it into Excel and change the FIPS code column from numeric to text for the FIPS codes to display properly. Nor can you double-click the file to open it in Excel and change the column type once you are in there either. It will not work. *You must follow these steps.*

1. **Go to your computer outside of ArcMap, and open Excel.**

2. **Click the Data tab, and then click Get External Text > From Text. For older versions of Excel, simply click the From Text button.**

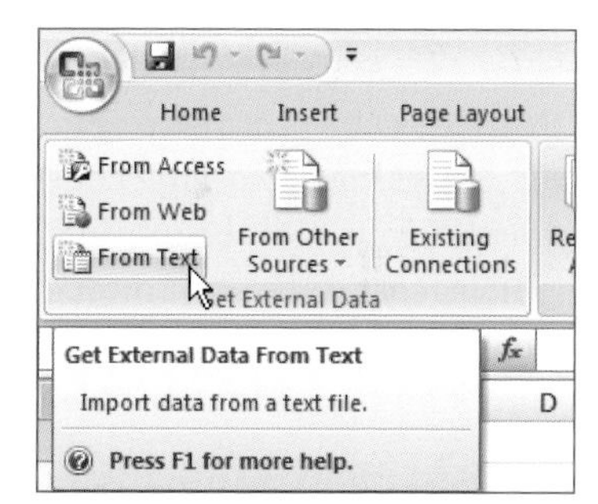

3. **Navigate to your save folder, and double-click CS_15_5YR_S0101_with_ann.csv.**

4. **On the Text Import Wizard dialog box, click Delimited as the original data type. Click Next.**

5. **In step 2 of the wizard, in the Delimiters panel, clear the Tab box and select the Comma box. Click Next.**

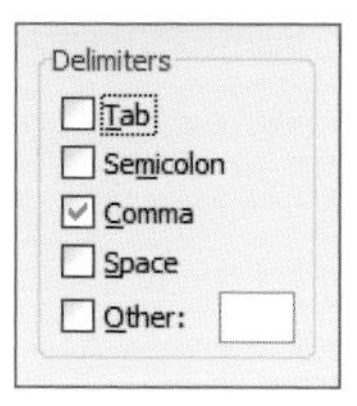

6. **In step 3 of the wizard, in the Data preview panel, is where the magic happens. Select the second column, Geo.id2, and then select the Text option in the Column data format panel to change that column type to text.** This text is the FIPS code, and now you'll learn a secret that most Excel users do not know: the FIPS code column is a text-based column, not a numeric column.

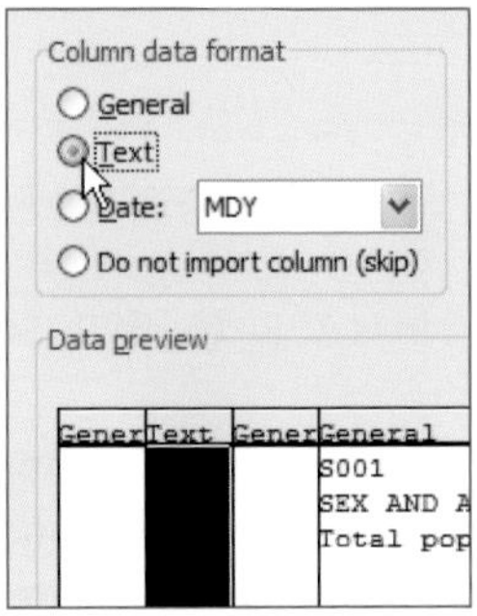

Column data format
General
Text
Date: MDY
Do not import column (skip)
Data preview

Gener	Text	Gener	General
			S001
			SEX AND A
			Total pop

7. **Click Finish, and then click OK.** Your data should appear in Excel.

Notice the FIPS code

The FIPS code provides a unique ID for every parcel of land in the US.

States have two-digit codes, counties three, and so a state plus a county code is a five-digit unique identifier for every county in the US. Nine states have a two-digit FIPS code that begins with zero. The whole reason you went through the trouble of importing data into Excel, instead of just opening it, is so you could keep the leading zero on the county five-digit code for these states. If this was not done, nine states would not be able to join.

1. **Confirm now that the second column, GEO.id2, has five digits. For example, the FIPS code for Autauga County, Alabama, should be 01001. If it is not, and displays as 1001, you missed doing the steps for importing into Excel correctly. You must start over.**

2. **If the data did not import correctly, close the table, and reimport it following steps 1 to 6 in the previous section, "Import into Excel."**

Prepping data for ArcMap

The purpose of preparing data in Excel first, instead of in ArcMap, is to simplify the data, focus on what you want to join (and eventually, thematically map), and create the optimal spreadsheet to import to ArcMap.

When downloading census data, you get numerous columns of irrelevant information. As you've done for this census data, it is helpful to decide early on exactly what you would ultimately like to display on your map. In chapters 5 and 6, you will join and create a thematic map using senior population. Although you've cleaned up the data before you were ever in Excel, there's a little more cleanup to do.

Clean up the spreadsheet

1. **Delete the first column, GEO.id, because it is unnecessary. (Do not delete your precious Geo.id2 column, which is the FIPS code and what you'll use to join this data to a map later on. Seriously, do not delete it.)**

In this exercise, you are mapping the senior population (percentage of population 65 years and older). Luckily, the Census Bureau provides this column already, and luckily, you correctly exported it earlier, which means you can sit back and relax and only do minor cleanup.

2. **Delete all rows at the top of the spreadsheet, leaving only one row for column headers. To join this table to a shapefile, you can have only one row at the top with column-heading information. Change the column headings to** ID**,** County**,** Population**, and** Seniors**.**

TIP *Once you have all the columns highlighted, simply clicking the Delete key on your keyboard would delete the data but not the columns. Right-clicking within the highlighted area and clicking Delete saves a step by deleting the columns.*

Clean up the county column

The County column text reads Autauga County, Alabama, but for labeling purposes later, it would be helpful to have only the county name with no additional text. To remove the state from the county column, use Excel's Find & Replace function.

1. **On the Home tab, click the Find & Select button, and click Replace.**

2. **In the Find what field, type** County, Alabama **(except use your state, of course). Type the label exactly as you see it for your state.**

3. **You will replace the name of the state with no text, so don't type anything in the Replace with field. Click Replace All.** And voila, just like magic, the tool deletes a bunch of useless text.

   Once you're finished with cleanup, the column should look like it does in the spreadsheet in the figure.

ID	County	Population	Seniors
01001	Autauga	55221	13.5
01003	Baldwin	195121	18.1
01005	Barbour	26932	16.1
01007	Bibb	22604	14.4
01009	Blount	57710	16.5
01011	Bullock	10678	14.6
01013	Butler	20354	17.8
01015	Calhoun	116648	15.5
01017	Chambers	34079	18
01019	Cherokee	26008	19.8
01021	Chilton	43819	14.9
01023	Choctaw	13395	20.3
01025	Clarke	25070	17.5
01027	Clay	13537	18.9
01029	Cleburne	15002	17.7
01031	Coffee	50884	15.4
01033	Colbert	54444	18.1
01035	Conecuh	12865	19.6
01037	Coosa	11027	17.9
01039	Covington	37886	19.6
01041	Crenshaw	13938	17.1
01043	Cullman	80965	17.3
01045	Dale	49866	14.9

Rename the worksheet

Naming Excel worksheets helps you stay organized. Perform the following step to name your worksheet.

1. **In the lower-left corner, double-click the Sheet1 tab (it may be Sheet0, depending on the version of Excel you are using), and type** AGE**.** You can bring in multiple worksheets from the same Excel workbook.

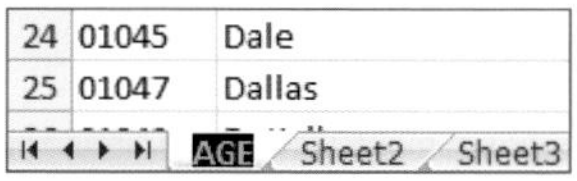

Save

1. **Click cell A1. Make sure nothing is highlighted.**

2. **Click the Office button (upper left) or click File, and click Save As. Navigate to your save folder. Rename the file** age.xlsx **(.xls is fine, too).**

3. **Close Excel.** You will not be able to open this spreadsheet in ArcMap if it is still open in Excel. In fact, you will likely get an error and have to reboot ArcMap, so it's important to fully close Excel.

4. **Take the rest of the day off.**

FILE NAMING: AVOID SPACES
Avoid spaces when naming files in ArcGIS. This rule is somewhat universal with file naming in general, but in ArcMap things can quickly come to a grinding halt over a space in a file name.

PERCENTAGES

It is a common task to map a percentage column. In Excel, you can change the column type to percentage. However, in ArcMap, this formatting will not be maintained. It is often necessary to change the column type to a percentage.

1. In ArcMap, add the table. Right-click the spreadsheet name in the table of contents, and click Open.
2. Right-click the column heading for the Percentage column (or for any column for which you want to change the column type), and click Properties.
3. In the Display panel, click the Ellipsis button ... next to Numeric.
4. In the Category panel, click Percentage. On the right, select "The number represents a fraction. Adjust it to show a percentage." Click the Numeric Options button.

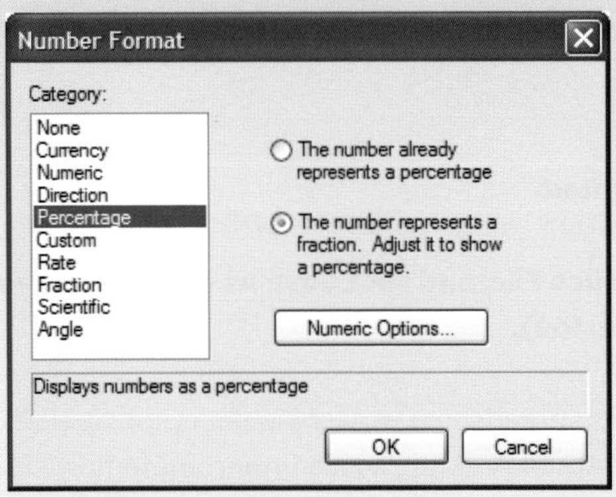

Number Format
Category:
None
Currency
Numeric
Direction
Percentage
Custom
Rate
Fraction
Scientific
Angle
The number already represents a percentage
The number represents a fraction. Adjust it to show a percentage.
Numeric Options...
Displays numbers as a percentage
OK
Cancel

5. Change the number of decimal places from 6 to 2. Click OK three times. Notice how the column has been reformatted.

These changes are not permanent. You must save the spreadsheet as part of saving an overall ArcMap project or as a layer (.lyr) file. However, if you use this spreadsheet again in another project, it will be necessary to make these changes again.

You can format the percentage in Excel ahead of time, not by changing the column type, but by formula, where you multiply the ratio by 100, resulting in a percentage such as 10.123. But then no % symbol is visible in the resulting ArcMap table. The % symbol is helpful to have in the column. When you build a legend for the resulting table, if the data contains a % symbol, it isn't necessary to type the word "Percent" to indicate the type of data. It will be apparent by the % symbol. Pros and cons exist for both approaches.

CHAPTER 5

KEY CONCEPTS

understanding unique IDs
understanding FIPS codes
joining Microsoft Excel files to maps

Joining data to maps

One of the most frequently used GIS skills involves connecting an Excel spreadsheet to a shapefile. This connection is where the magic of GIS happens. Often, the purpose of joining data to a map is to visually display the distribution of a dataset through a thematic map (covered in chapter 6). Joining your own data to a shapefile can be extremely useful.

It's hard to overemphasize how important this skill is. It's somewhere between oxygen and pizza. In this exercise, you will learn how to join your own data to a map. This ability to join data to maps is straight-up amazing and a precursor to some fantastic thematic mapping.

You will need to download all chapter data files from esri.com/GIS20-3.

Add two files to a join

1. **Open ArcMap. Click Add Data, and add countiesprj.shp from the chapter 3 folder.**

2. **Click Add Data, and add age.xlsx by double-clicking the file name and then Age$.** If you did not change the name of the worksheet in the last exercise in chapter 4, the existing worksheet will be called Sheet1$ or Sheet0$. Worksheets are denoted with $ in the name.

Double-check and find the FIPS columns

1. **Check to make sure the data is correct. The Age$ data table should now appear in the table of contents. To view the data table, right-click the data table name, and then click Open. Review the data to make sure it looks as you would expect it to look.**

   NOTE: *The attribute table has a way of taking up the full screen. If it does, don't panic. Simply grab the edges of the attribute box and make it smaller.*

   This next step is superimportant. To join data to maps, you must link two columns that have overlapping data, one column from the data table and its comparable column in the map layer.

2. **Identify the two columns you will use for joining by opening the attribute table for each column.** The Age$ table is already open.

3. **In the table of contents, right-click the county shapefile, and then click Open Attribute Table. Notice two tabs are now open at the bottom of the table.**

4. **Click each tab in the lower-left corner, review each one, and find two columns that match. The column names do not have to be the same, but the content of the columns does.** In this example, the column name in the shapefile attribute table is GeoID, and the column name in the spreadsheet is ID.

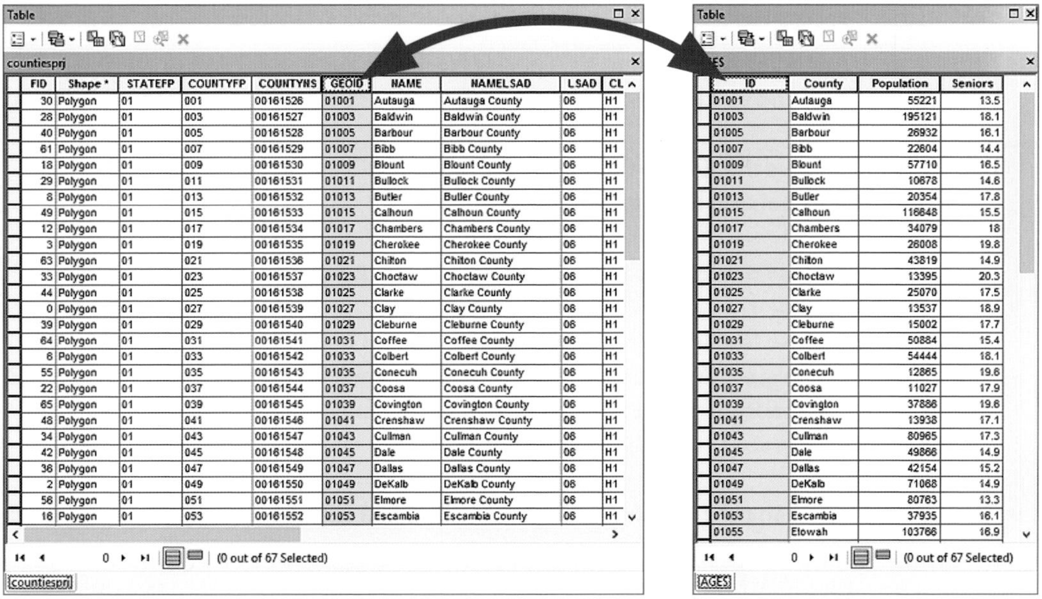

Table

countiesprj

FID	Shape *	STATEFP	COUNTYFP	COUNTYNS	GEOID	NAME	NAMELSAD	LSAD	CL
30	Polygon	01	001	00161526	01001	Autauga	Autauga County	06	H1
28	Polygon	01	003	00161527	01003	Baldwin	Baldwin County	06	H1
40	Polygon	01	005	00161528	01005	Barbour	Barbour County	06	H1
61	Polygon	01	007	00161529	01007	Bibb	Bibb County	06	H1
18	Polygon	01	009	00161530	01009	Blount	Blount County	06	H1
29	Polygon	01	011	00161531	01011	Bullock	Bullock County	06	H1
8	Polygon	01	013	00161532	01013	Butler	Butler County	06	H1
49	Polygon	01	015	00161533	01015	Calhoun	Calhoun County	06	H1
12	Polygon	01	017	00161534	01017	Chambers	Chambers County	06	H1
3	Polygon	01	019	00161535	01019	Cherokee	Cherokee County	06	H1
63	Polygon	01	021	00161536	01021	Chilton	Chilton County	06	H1
33	Polygon	01	023	00161537	01023	Choctaw	Choctaw County	06	H1
44	Polygon	01	025	00161538	01025	Clarke	Clarke County	06	H1
0	Polygon	01	027	00161539	01027	Clay	Clay County	06	H1
39	Polygon	01	029	00161540	01029	Cleburne	Cleburne County	06	H1
64	Polygon	01	031	00161541	01031	Coffee	Coffee County	06	H1
6	Polygon	01	033	00161542	01033	Colbert	Colbert County	06	H1
55	Polygon	01	035	00161543	01035	Conecuh	Conecuh County	06	H1
22	Polygon	01	037	00161544	01037	Coosa	Coosa County	06	H1
65	Polygon	01	039	00161545	01039	Covington	Covington County	06	H1
48	Polygon	01	041	00161546	01041	Crenshaw	Crenshaw County	06	H1
34	Polygon	01	043	00161547	01043	Cullman	Cullman County	06	H1
42	Polygon	01	045	00161548	01045	Dale	Dale County	06	H1
36	Polygon	01	047	00161549	01047	Dallas	Dallas County	06	H1
2	Polygon	01	049	00161550	01049	DeKalb	DeKalb County	06	H1
56	Polygon	01	051	00161551	01051	Elmore	Elmore County	06	H1
16	Polygon	01	053	00161552	01053	Escambia	Escambia County	06	H1

0 (0 out of 67 Selected)

countiesprj

Table

ID	County	Population	Seniors
01001	Autauga	55221	13.5
01003	Baldwin	195121	18.1
01005	Barbour	26932	16.1
01007	Bibb	22604	14.4
01009	Blount	57710	16.5
01011	Bullock	10678	14.6
01013	Butler	20354	17.8
01015	Calhoun	116848	15.5
01017	Chambers	34079	18
01019	Cherokee	26008	19.8
01021	Chilton	43819	14.9
01023	Choctaw	13395	20.3
01025	Clarke	25070	17.5
01027	Clay	13537	18.9
01029	Cleburne	15002	17.7
01031	Coffee	50884	15.4
01033	Colbert	54444	18.1
01035	Conecuh	12865	19.6
01037	Coosa	11027	17.9
01039	Covington	37888	19.6
01041	Crenshaw	13938	17.1
01043	Cullman	80965	17.3
01045	Dale	49866	14.9
01047	Dallas	42154	15.2
01049	DeKalb	71068	14.9
01051	Elmore	80763	13.3
01053	Escambia	37935	16.1
01055	Etowah	103766	16.9

0 (0 out of 67 Selected)

AGES

It is imperative that you understand the concept here. You have two columns, and you are going to link the map to the data table using these columns. They contain identical information. Note the five-digit FIPS code in each one. You can even sort them so you can compare them line by line.

TIP ***If the join does not work, go back and click Validate Join, which will give you clues about why it didn't work.***

5. **Close the attribute tables.**

Join the data table to a map

1. **In the table of contents, right-click the countiesprj.shp shapefile (not the data table from Excel).**

2. **Click Joins and Relates, and then click Join.**

3. **In the "What do you want to join to this layer?" field, select "Join attributes from a table."**

4. **In the "Choose the field in this layer that the join will be based on" field, select the appropriate column heading—in this case, GeoID.**

5. **In the "Choose the table to join to this layer" field, Age$ will already be selected.**

6. **In the "Choose the field in the table to base the join on" field, ID will already be populated.** If ID is not populated, the column is incorrectly formatted. The genesis of this error is not importing the Excel spreadsheet into Excel, but rather just opening it. You will need to close ArcMap and Excel, and go back to chapter 4, "Import into Excel."

7. **Select the Keep Only Matching Records option, and click OK. To save doing an extra step, skip the Validate Join button.**

Verify the join worked correctly

1. **Right-click the shapefile name, and then click Open Attribute Table.**

2. **Scroll to the far right to see if data from the spreadsheet has been appended to the end of the attribute table.** You should not see any error messages or null values.

INTPTLON	ID	County	Population	Seniors
085.8635254	01027	Clay	13537	18.9%
087.7910910	01091	Marengo	20306	18.4%
085.8039920	01049	DeKalb	71068	14.9%
085.6542417	01019	Cherokee	26008	19.8%
087.6230608	01065	Hale	15256	17%
087.2938269	01105	Perry	10038	17.6%
087.8014569	01033	Colbert	54444	18.1%
087.9642005	01063	Greene	8697	17.4%
086.6819689	01013	Butler	20354	17.8%
085.3530477	01081	Lee	150982	9.9%
088.1965682	01097	Mobile	414251	14.1%
087.7642923	01057	Fayette	16896	19%
085.3940321	01017	Chambers	34079	18%
087.5227834	01125	Tuscaloosa	200458	11.4%
087.3049075	01131	Wilcox	11235	17.1%
086.3216681	01095	Marshall	94318	16%
087.1684097	01053	Escambia	37935	16.1%
086.9813995	01083	Limestone	88805	13.4%

3. **Double-check the number of records in the Age$ tab by right-clicking the file and clicking Open.** In the lower-right corner, the number of records is listed (in this case, 67).

4. **Now check the number of records in the newly joined shapefile—it should be the same number. If it is not, the two columns are not identical and must be corrected.**

5. **Close the attribute table.**

Create a new shapefile

When files are joined, it is a temporary join. To permanently join these files, you will use a process to essentially create a copy of the joined file. The process cements the join into one file. The file you're about to create will permanently warehouse the data from the age file and can be used in other projects.

1. **In the table of contents, right-click the shapefile name, click Data, and then click Export Data.**

2. **Click the Browse button to browse to your save folder, and name the new shapefile** agejoined **(no spaces in file names). Ensure that "Save as type" is shapefile. Click Save. Verify that it is saving where you want it to save, and then click OK.**

3. **When asked if you want to add the exported data to the map as a layer, click Yes. Notice the new file added to the table of contents.**

4. **To remove the original Excel file and shapefile, which are no longer needed, right-click the Age$ file, and click Remove. Then right-click the original county file, and click Remove.**

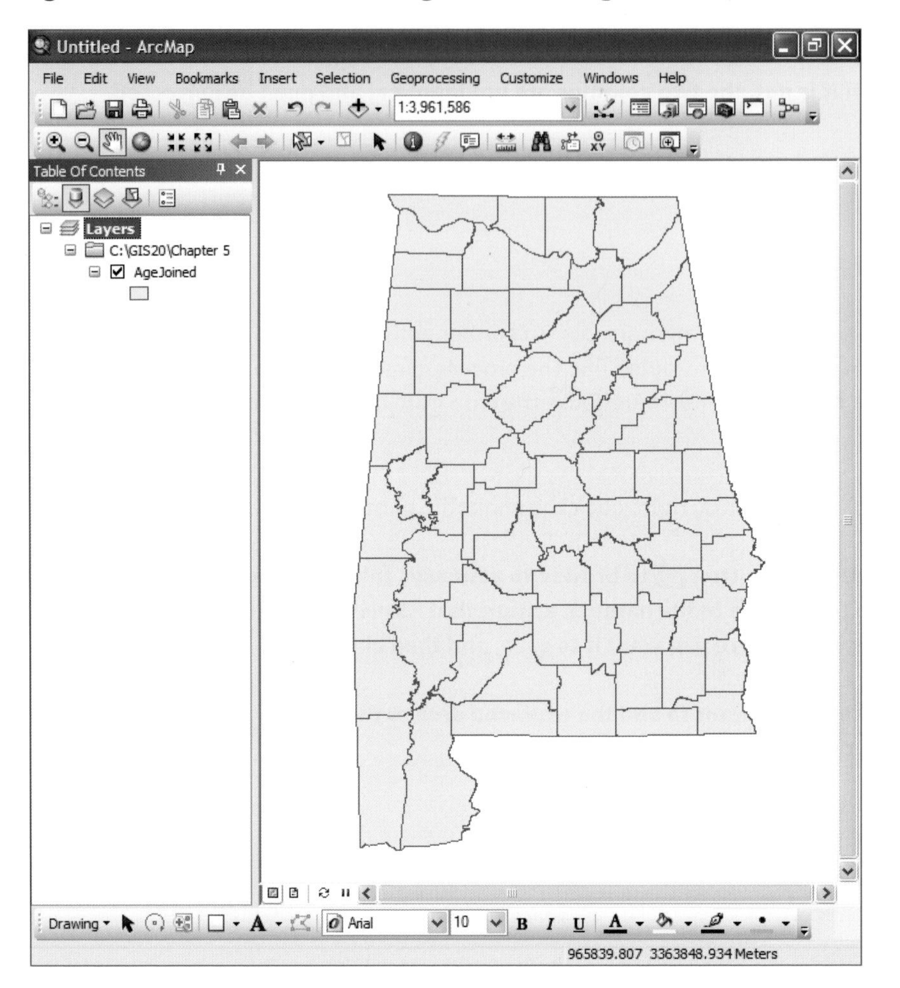

Congratulations! You now have a permanently joined shapefile (agejoined) that contains data about the senior population. You will use this file in the next chapter.

CHAPTER 6

KEY CONCEPTS

changing column types
using color ramps
creating custom legends
exploring layouts

Creating thematic maps

Thematic mapping enables you to show distribution of data across geography. It is one of the most frequently used GIS tools.

In this exercise, you will create a thematic map. This mapping is the kind that rock stars do. You have to be careful, however—as soon as people know you can do it, there will be a line around the block for your mapping services.

You will need to download all chapter data files from esri.com/GIS20-3.

Create a thematic map

1. **The shapefile agejoined should still be open from the last exercise. If not, use the Add Data button ✚▾ to add this file to ArcMap.**

2. **In the table of contents, right-click the layer name, and click Properties. Click the Symbology tab.**

3. **On the left, select Quantities > Graduated Colors.**

4. **In the Value list, click the column of data you want to map—in this case, Seniors. And on the color ramp, select the first color ramp. Click OK.**

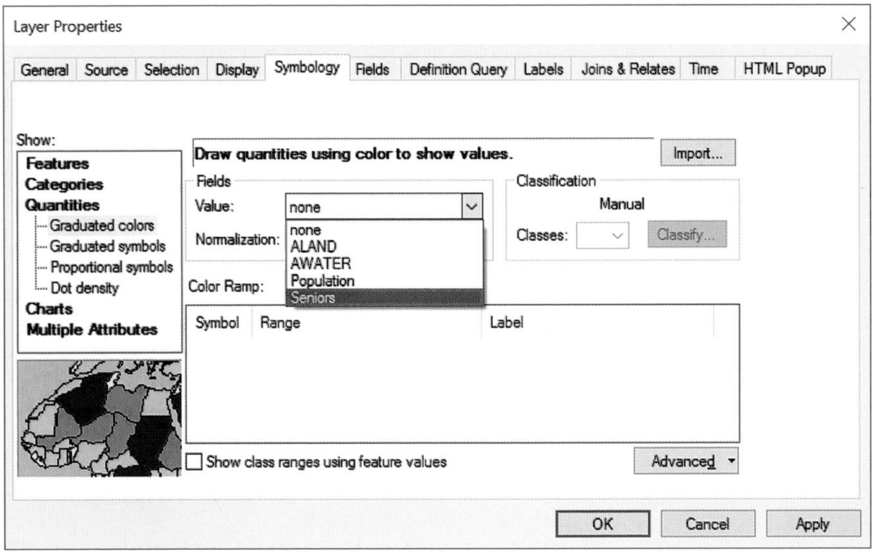

Change the color ramp

Okay, the choice of colors looks kind of crazy. So now you can fix the colors.

1. **In the table of contents, right-click the layer name, click Properties, and then click the Symbology tab.** Monochromatic blue is a good choice for thematic maps because it looks professional, and readers can easily distinguish between the shades of blue. Grays, tans, and greens are also good choices. Avoid reds, oranges, pinks, and purples. Color ramps don't have names.

2. **Select the monochromatic blue color ramp. Click OK.**

Color Ramp:

Fix the percentages

The percentages display in an obnoxious way in the table of contents. They would also display this way in the legend. So now you can fix how they look.

1. **In the table of contents, right-click the layer name, and click Open Attribute Table.**

2. **Scroll far right, and look at the Seniors column.** The values are displayed with one decimal place (e.g., 9.1), but in the table of contents, the values are displayed with six decimal places (e.g., 9.100000).

3. **Right-click the Seniors column heading, and click Properties.**

4. **Click the Ellipsis button [...] on the right of the word Numeric.**

5. **In the Category panel, click Percentage. Make sure "The number already represents a percentage" is selected, and then click Numeric Options.**

6. **Change the number of decimal places from 6 to 1. Click OK three times. Notice that now there is a % symbol at the end of the percentages, which will look good in the legend. Close the attribute table.**

   You need a way to refresh the values in the table of contents.

7. **Right-click agejoined, and click Properties. Select the Seniors column from the Value list again. Click OK.** This step refreshes the table of contents and displays the percentage in a reader-friendly way.

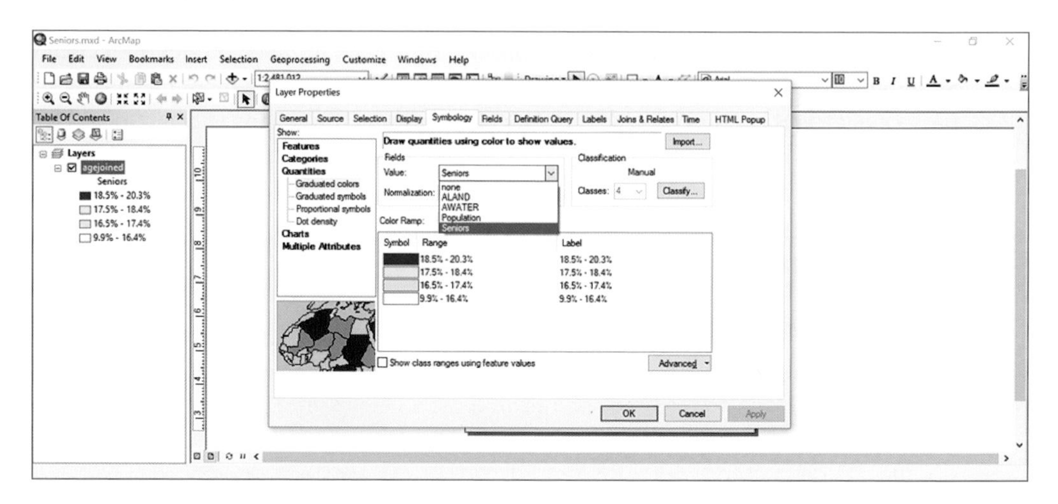

Change the way the legend breaks

The natural breaks method for thematic maps is suitable for informal mapping; however, for a more sophisticated approach, you should manually break the legend. Rather than deciding where to break it randomly, you'll use the average for your dataset as the first break point, thereby establishing a baseline for normalcy.

Classes are the data ranges that appear in a map's legend.

The default is five. Four intervals may be preferable for a general, nontechnical audience.

1. **In the table of contents, right-click the layer name, click Properties, and click the Symbology tab.**

2. **In the Classes box, change the number of classes from 5 to 4, and click Classify.**

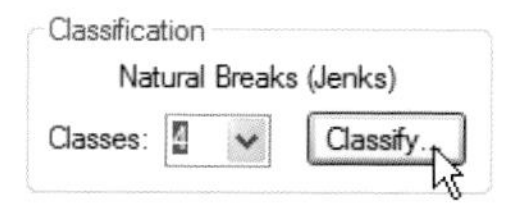

3. **Look in the statistics box (upper right) to determine the mean. Write down that number. In this example, it is 16.4 percent.**

4. **To change the first break point to the mean of your data (in this case, 16.4), manually type over the given numbers in the Break Values section.** For Alabama, the data distribution only goes to 20.3 percent, so the break points after the 16.4 percent must be very small increments.

5. **Change the second and third break points to something that makes sense for your data range. Never change the last break point, as this is the natural end of your data. Here, use** 17.4 **and** 18.4**, and leave the last break point at 20.3. Click OK twice, and look at the map. Notice you can more easily see a pattern emerging from the data.**

Place the highest values at the top

To create a map that emphasizes the concentration of a variable (versus the lack of a variable), construct the legend so the highest values are at the top. The legend has not been created yet, but the content of the table of contents will be the content of the legend. Sort the classes so the highest values are at the top.

1. **In the table of contents, right-click the layer name, click Properties, and click the Symbology tab.**

2. **Underneath the color ramp, notice the headings Symbol, Range, and Label. Click the Range column heading, and select Reverse Sorting.**

3. **Click the Symbol column heading, and select Flip symbols.**

Change the color (again)

Now that you have applied color to your map, a pattern may be emerging. You may be able to make this pattern even clearer. You can de-emphasize less interesting areas by not shading them with a color, and emphasize areas of interest by applying color. The less color on your map, the easier it will be to read. Using white to denote the lesser values in the map will make a pattern easier to see and look more professional.

Change the map colors to emphasize areas with high proportions of seniors.

1. **In the Symbol column, notice the four color swatches. Change the lowest range to white (not hollow). Simply click the color, and change the fill color to white. Next, change the second range to a light gray. Change the top two classes with the highest values to medium and dark blue, with dark blue signifying the highest values. Click OK, and look at your map.**

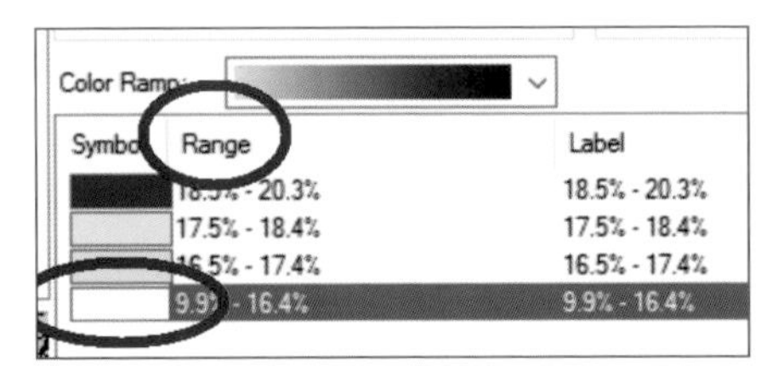

This symbology choice de-emphasizes low data values, while emphasizing high data values with color.

Label the counties

1. **Turn on labels for the counties by right-clicking agejoined, clicking Properties, clicking the Labels tab, and selecting the "Label features in this layer" check box in the upper-left corner. Click OK.** Applying a halo will help the label stand out.

Create a layout

1. **On the View menu, click Layout View. Create a layout that includes a legend, title, source, north arrow, and scale bar. For a review of these steps, see chapter 2. Use the Magnifier and the Pan tool to make your state big and centered.** Psst! Everything you need is on the Insert tab on the menu bar.

> **THEMATIC MAP TITLES**
> When deciding what to call a thematic map, consider beginning the title with the phrase "Distribution of ..."—for example, "Distribution of Poverty Rates by State, 2010." This phrasing works well with many maps.

Fix the legend

Often the legend will be clearer by removing layer names and column headings. In this example, the legend doesn't really need the text "agejoined" and "Percent."

1. **To remove this text, double-click the legend to go directly to Legend Properties.**
2. **Click the Items tab.** There is a secret menu here accessible by right-clicking agejoined underneath the Select All and Select None buttons.
3. **Right-click agejoined, and then click Properties.**

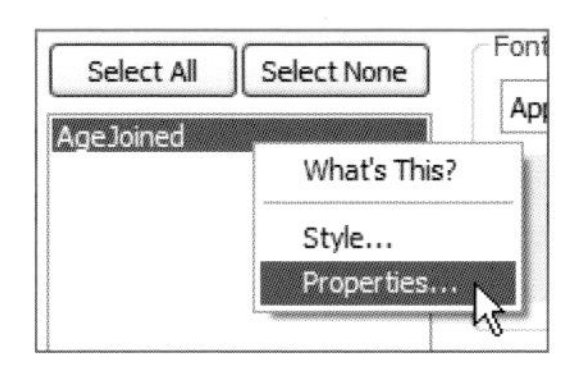

4. **Click the General tab, and clear the Show Layer Name and Show Heading check boxes. Click OK twice.**

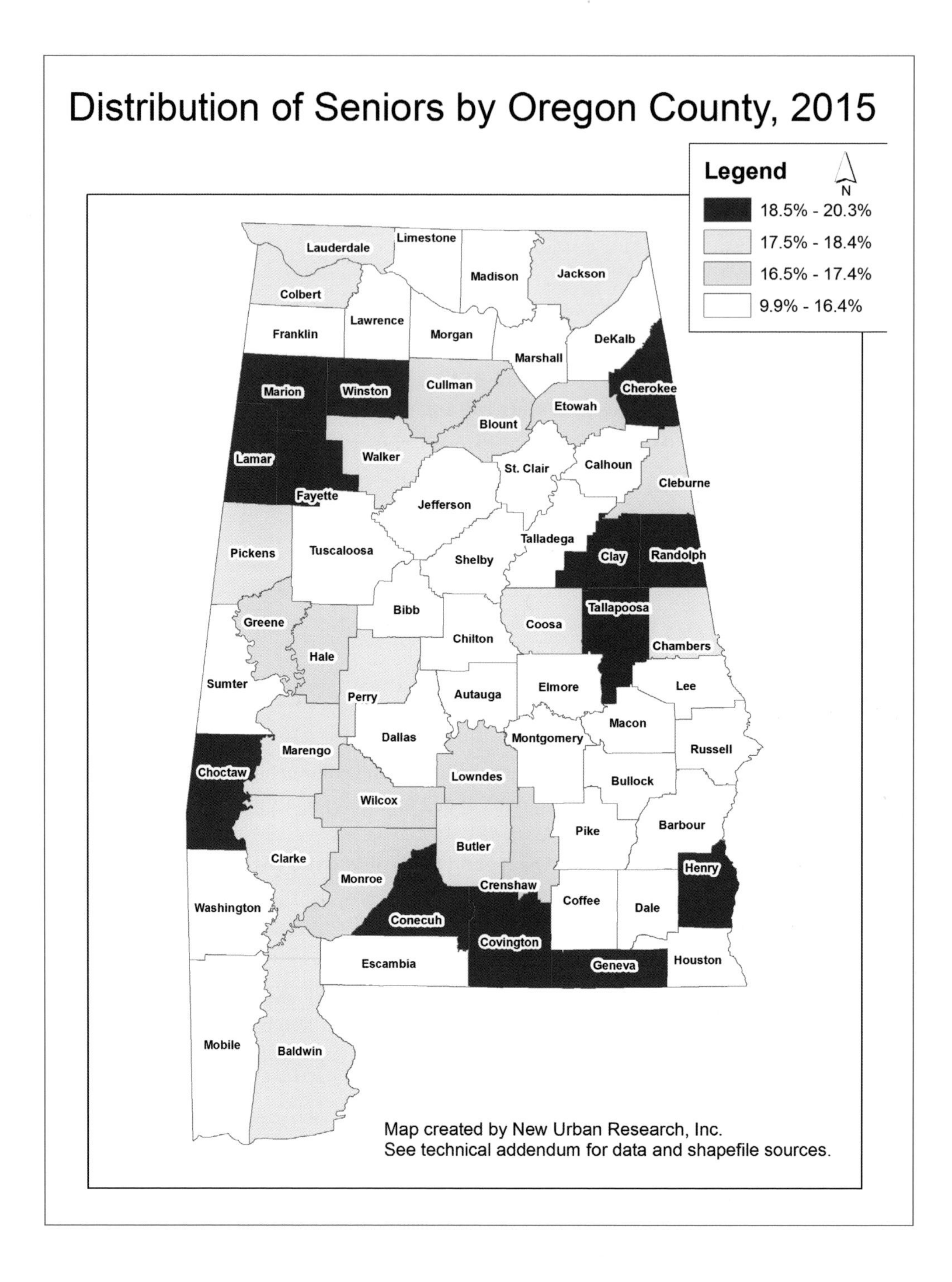
Distribution of Seniors by Oregon County, 2015
Legend
N
18.5% - 20.3%
17.5% - 18.4%
16.5% - 17.4%
9.9% - 16.4%
Lauderdale
Limestone
Madison
Jackson
Colbert
Lawrence
Franklin
Morgan
DeKalb
Marshall
Marion
Winston
Cullman
Cherokee
Etowah
Blount
Lamar
Walker
St. Clair
Calhoun
Cleburne
Fayette
Jefferson
Talladega
Pickens
Tuscaloosa
Shelby
Clay
Randolph
Bibb
Tallapoosa
Greene
Coosa
Chilton
Chambers
Hale
Sumter
Autauga
Elmore
Lee
Perry
Macon
Dallas
Montgomery
Russell
Marengo
Choctaw
Lowndes
Bullock
Wilcox
Barbour
Pike
Butler
Clarke
Henry
Crenshaw
Monroe
Coffee
Dale
Washington
Conecuh
Covington
Escambia
Geneva
Houston
Mobile
Baldwin
Map created by New Urban Research, Inc.
See technical addendum for data and shapefile sources.

LEGEND ISSUES

Here's something good to know: if inserting a border and background color of white is skipped the first time through, it will cause one of two problems.

PROBLEM 1: NO ROOM BETWEEN THE BORDER AND EVERYTHING ELSE

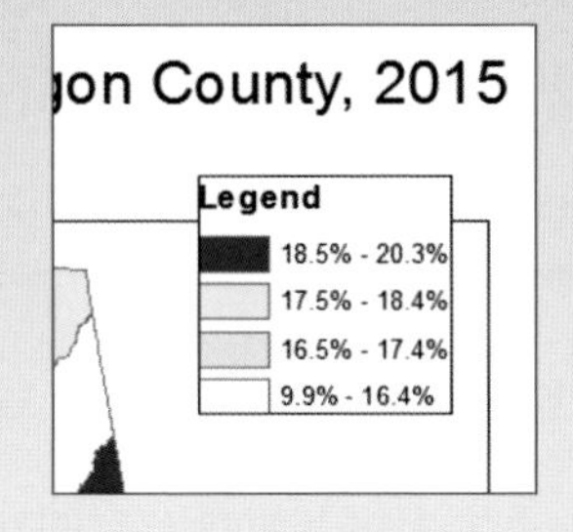

PROBLEM 2: TRANSPARENT GAP BETWEEN THE BORDER AND EVERYTHING ELSE

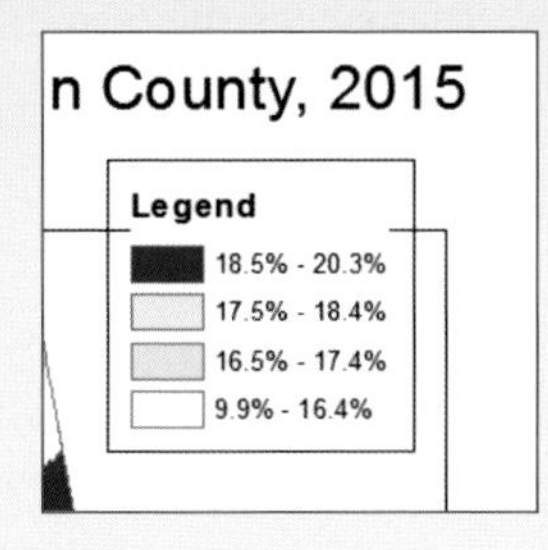

Solution: delete the legend by clicking on it and clicking Delete on the keyboard. Insert the legend again, but this time, the first time through on the Frame tab, select a border and a background color of white. If you insert the legend and forget to do all this, and then go back and try to fix it, you'll get problem 1.

If you insert the legend but remember to insert a border, but then forget to insert the background color of white (or misunderstand and think it is already white when it's not—it's transparent), and then go back and try to put in a background color of white, you'll get problem 2. The only way to get the legend right is to assign a border and background color of white the first time through.

Save the project

1. **On the File menu, click Save As, and navigate to your save folder.**

2. **Name this file** Seniors.mxd **and close ArcMap, or leave it open for the next exercise.**

CITING THE CENSUS BUREAU: APA WEB CITATION

Maps must include information on the sources of data and shapefiles used, as well as the map creator. Here is one way to cite Census Bureau data and shapefiles:

- Map source: New Urban Research Inc. 2017.
- Data source: 2015 American Community Survey, Table S0101Sex by Age, ACS 2015.
- 5-year estimates. Retrieved July 17th, 2017, from https://factfinder.census.gov/faces/tableservices/jsf/pages/productview.xhtml?pid=ACS_15_5YR_S0101&prodType=table.
- Shapefile source: US Census Bureau (2016, Jan. 1), 2016 TIGER/Line Shapefiles for: Alabama, retrieved July 17, 2017, from US Census Bureau: http://www.census.gov/cgi-bin/geo/shapefiles/index.php.

CHAPTER 7

KEY CONCEPTS

adding/deleting columns
editing values
making calculations

Working with data tables

Manipulating data tables (attribute tables) includes adding and deleting columns, editing values, and performing calculations. Although the bulk of data manipulation is best achieved outside ArcMap, it is important to learn the fundamentals of data manipulation within ArcMap. Editing the data will greatly strengthen your analysis and improve your maps.

In this exercise, you will learn to edit data tables. Not flashy, but a solid skill to have.

[*You will need to download all chapter data files from esri.com/GIS20-3.*

Add a shapefile, and open the attribute table

1. **Click Add Data ▾ to add agejoined.shp created in chapter 5, also located in the chapter 7 folder of downloaded data from the book resource web page.**

2. **Open the attribute table by right-clicking the layer name in the table of contents and clicking Open Attribute Table.** You should see data in spreadsheet format.

Edit existing data in an attribute table

If you try to edit any of the data in the table, you will notice the values cannot be changed. You must first make the table editable.

1. **Move the attribute table out of the way. Click the Editor Toolbar button .** This tool is also available by going to Customize > Toolbars > Editor. Either way, a new Editor toolbar will be visible.

2. **If you have an older version of the software, go to the View tab, click Toolbars, and then click Editor.**

3. **Click the Editor button Editor ▾ , and then click Start Editing. Notice the top row of the attribute table turns white, which indicates that the table is now editable.**

4. **Scroll left and find the ClassFP column. Type your new values over any of the values in this column.**

5. **When you are finished, click Editor again, and then click Stop Editing.**

6. **Save edits when prompted, and close the attribute table.**

Edit data outside the attribute table on a polygon-by-polygon basis

You can also edit data within an information box instead of doing it within the attribute table. One advantage of this method is that you can click individual polygons and edit data for that polygon.

1. **On the Editor toolbar, click the Editor button, and then click Start Editing.**
2. **On the Editor toolbar, click the Attributes button.**
3. **On the map, click any county. Notice that a new window opens on the right with all the information from the underlying attribute table.**
4. **Click the ClassFP field, and replace H1 with** H3 **for the value in this field.**
5. **On the Editor toolbar, click Editor > Save Edits > Stop Editing.**
6. **Close the Editor toolbar by right-clicking the toolbar and clicking Editor.**

Add a column to the attribute table

1. **Open the agejoined attribute table. On the Table toolbar, click the Table Options button, and click Add Field.**

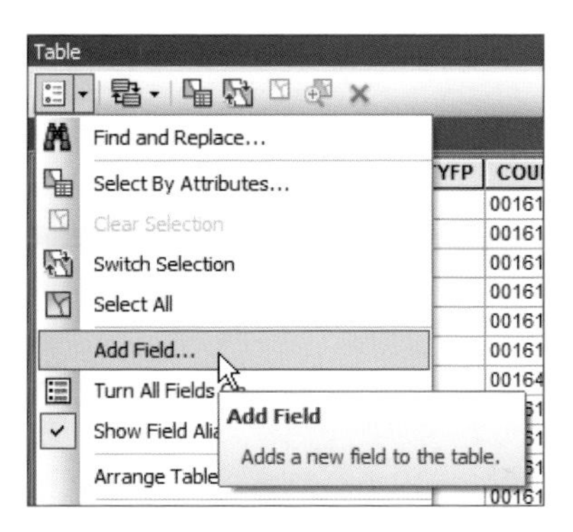

NUMERIC COLUMN TYPES

Generally, you should use float or double (also known as a double float) as the column type. The double float is beneficial because you can have the maximum number of digits for the number, with no rounding (after 15 digits, plain old floats start to round, but doubles do not).

2. **Type the name of the first new column. Call it** Percent. **For Type, select Float.** Float is a flexible column type so it is often used. Float (as well as double) is useful because the number can be a decimal or whole number.

3. **Leave the Precision and Scale boxes as zeros for the maximum digits. Click OK.**

4. **Repeat steps 1 to 3, adding a second column named** Count. **Select the same options for the new field.** The columns are added to the end of the attribute table.

   NOTE: *If Add Field is dimmed, it means you are still in an edit session. You must end the session, and try again.*

Make calculations

Calculations can be performed in or out of an edit session. Now you can derive the number of senior citizens in each county.

First, you must get the percentage of seniors, contained in the Seniors field, into a format that, when multiplied by total population, will return the number of senior citizens in each county.

1. **Right-click the Percent column header, click Field Calculator, and accept the warning.** If you don't get a warning, that's fine, too. The formula to transform the percentage into a usable one for your purposes is Seniors / 100.

2. **Double-click Seniors from the list of variables, click the / button (the division symbol), and then type** 100. **Click OK.** The formula is automatically filled in as you go: [Seniors] / 100 . It will calculate the percentage.

   Now you're ready to derive the percentage of seniors per county.

3. **Right-click the column heading for the Count column, and click Field Calculator.**

   Now you're ready to derive the count of seniors.

4. **Erase the old formula, and double-click Population from the list of fields. Click * (the multiplication symbol), and click the Percent column. Click OK to calculate the count.**

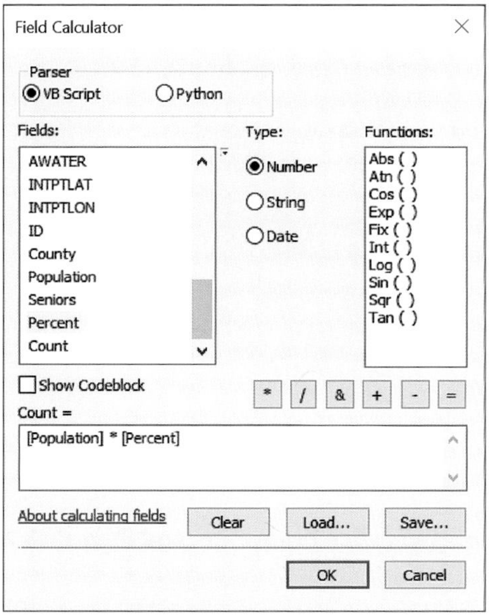

5. **Right-click the Count column heading, click Properties, and click the ellipsis button ... . Change the number of decimal places from 6 to 0, and select the check box to show thousands separators. Click OK twice.** This step will transform this column into the correct format.

6. **Sort the Seniors column heading (descending) to see which counties had the highest percentage of seniors, and notice how that relates to the count field.** Of all counties in Alabama, Choctaw County has the highest percentage of senior population, even though the county has only 2,719 seniors overall. Interesting.

Delete columns

You may notice several prebuilt columns when you download shapefiles from the Census Bureau. Some columns are useful for understanding what geography is within the shapefile, but you can delete many of these columns because they are rarely used. You can delete columns two ways. The method you use depends on how many columns of data you want to delete. If it is only one or a few columns, you would use the following method (for several columns, a different method must be used).

To delete one column of data, do the following steps:

1. **Right-click the LSAD column heading, and click Delete Field.**

2. **Click Yes to delete the LSAD column.**

   To delete multiple columns of data, do the following steps:

3. **Close the attribute table.**

4. **In the table of contents, right-click the layer name, click Properties, and then click the Fields tab (center).**

5. **Click the Turn All Fields Off button, and notice that the check boxes next to all the columns are cleared.**

6. **Select only those columns you want to keep. The essential columns to keep are GeoID, Name, and Percent. Click OK.**

7. **In the table of contents, right-click the layer name, and open the attribute table. Notice that now only three columns are displayed.** The other columns are still there but hidden. Simply hiding the columns might work for some projects, but for many other projects, you might want only these three columns.

8. **Close the attribute table.**

   Now you can essentially make a copy of this file and, in so doing, capture only these three columns.

9. **In the table of contents, right-click the layer name, click Data, and then click Export Data.**

10. **Click the Browse button, and navigate to your save folder. Type a new name such as** minifile**, make sure the "Save as type" is Shapefile, and click Save.**

11. **Click OK and click Yes to add the file to the map as a layer.**

12. **Open the attribute table to verify you have only the columns you exported (plus a couple of others that are standard).**

Work with multiple attribute tables

1. **Open both attribute tables (agejoined and minifile).**

2. **Click each tab in the lower-left corner of the table's window, and notice how you can switch between tables.** The tabs can be moved by dragging and dropping.

01053	Escambia	15.2
01065	Hale	15
01131	Wilcox	15

Minifile AgeJoined

You can also dock the tables to create a split screen.

3. **Drag the minifile tab to the center of the table window. A blue circle with four arrows will appear. Drop the tab on the right arrow. The screen will split with one table on each side.**

4. **To undo the split screen, simply drag the table back into the tab position.**

5. **Close ArcMap, or leave it open for the next exercise.**

CHAPTER 8

KEY CONCEPTS

using address locators
auto and manual geocoding
using street networks and other layers

Address mapping

Address mapping, also called *geocoding*, is like creating a pushpin map of addresses. Geocoding is a skill everyone needs in their GIS work. You might use this skill for mapping diseases, crimes, client addresses, service addresses, or anything with a physical location—an address—you want to show on a map.

[*You will need to download all chapter data files from esri.com/GIS20-3.*

Geocoding is an amazing skill to have in your toolbox. Geocoding and thematic mapping are hands down the two most desirable skills in GIS. In this exercise, you'll learn how to do geocoding.

Geocoding in ArcMap has changed a lot over the years. In 2012, Esri® released ArcMap 10.1, which introduced new geocoding technology. Previously, users created their own address locators (the thing that makes geocoding work), but it was a lot of work. Since 2012, Esri has hosted an address locator on its servers upstream, and ArcMap connects to it in the background as you work. The Esri-hosted locator not only geocodes using street network files, but also by ZIP Codes and cities, as well as using a US Postal Service Master Address File, which is unavailable to the average person. This address locator makes geocoding a breeze. The problem is, it costs money to use it.

Distribution of Select Social Services
Bexar County, Texas 2017

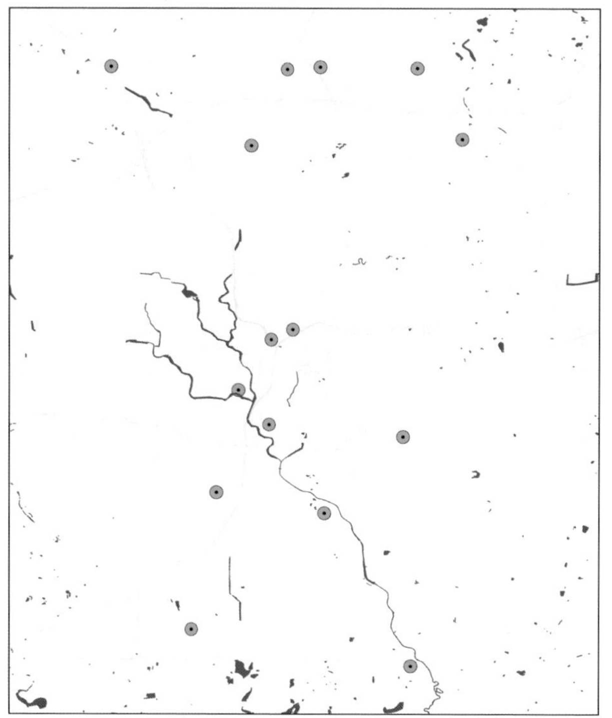

Geocoding is now a fee-for-service, in which users buy credits and use them to geocode. Esri provides a way to get 200 free credits, which will allow you to geocode about 4,000 addresses. In this chapter, you'll get more than enough credits to complete the exercise and geocode some addresses. If you want to go old school (and free), included at the end of the chapter are instructions on how to create your own address locator.

Geocoding using Esri's address locator (credits required)

As mentioned, to geocode in ArcMap, you need some credits. If you already have access to credits, such as through your organization, you do not need to set up an ArcGIS trial. The steps below assume you do not have access to geocoding credits. If you think you already have access to credits, you will need the login and password for the ArcGIS℠ Online organizational account. Then you can skip right to "Add List of Social Service Agencies" later in this section.

This part is a little confusing. What you are supposed to do first depends on whether you have credits. If you downloaded the ArcGIS® Desktop Advanced 180-Day Trial for Esri Press included with this book (but not with the e-book), you already have 200 free credits. All you need is your user name and password. If you did not download the software, perhaps because you already had it installed, you will need to sign up for the Esri-sponsored ArcGIS Trial, not for the software, but the credits. Follow these steps to set up an ArcGIS trial and get credits.

Set up ArcGIS Trial (for the purpose of getting credits)

1. **Navigate to http://www.esri.com/arcgis/trial.**

2. **Fill in your information to sign up for the ArcGIS Trial.**

   NOTE: *If you get a message saying that your email is already in use, use a different email. If you don't have any other email, it's simple to sign up for a free Gmail account.*

3. **Once you've completed the form, check your email, and click the activation link. It will ask you to establish a user name and password. It's a good idea to use your email as the user name, even though it's possible to create a user name that is not in email format. Once you accept the terms of use, click the Create My Account button.**

4. **A message will ask if you want to download the ArcGIS Trial, which you do not need to do. Instead, go to ArcGIS Online.**

5. **You will be prompted to fill out a profile. It does not matter what you input in these fields.** To know if you've successfully gotten those credits, you'll see in the upper-right corner of the Organization page, under Subscription Status, "200 credits remaining." Solid gold.

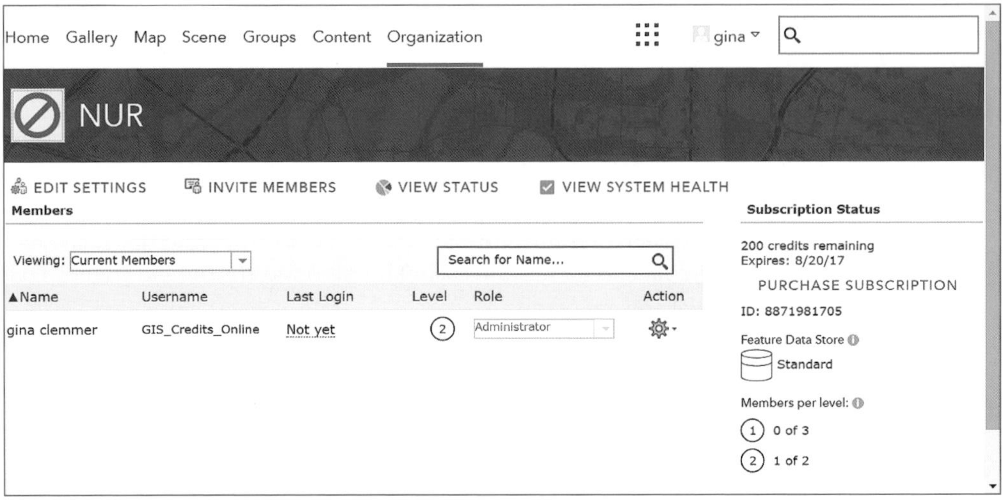

Add list of social service agencies

1. **Open ArcMap.** You have some addresses to geocode.

2. **Click Add Data to add the file Agencies.xls, Agencies$ worksheet.**

3. **In the table of contents, right-click Agencies$, and click Open. Have a look around.** I've only included 20 agencies in this spreadsheet to conserve your precious credits, since you might want to geocode addresses for real projects you're currently working on.

4. **Notice the Address field, as well as the City, State, and Zip (for ZIP Code) fields.** The Address field is the one you'll use for geocoding, but also important is that the corresponding fields are in the proper format for addresses. The street name—and the city, state, and ZIP Code information—should all be in their respective columns.

Table

Agencies$

Agencies	Address	City	State	Zip	Type
Barton House	10 Lynn Batts	San Antonio	TX	78218	AGED HOME
Bright Horizons	100 Citibank Drive	San Antonio	TX	78245	CHILD DAY CARE SERVICES
Alloy Wheel Repair Special	10002 Cedarbend Drive	San Antonio	TX	78245	HELPING HAND SERVICE (BIG BROTHER, ETC.)
Rosies Child Care Center	10002 Weybridge	San Antonio	TX	78250	CHILD DAY CARE SERVICES
U S Injury Rehabilitation Ctrs	10004 Wurzbach Road 361	San Antonio	TX	78230	VOCATIONAL REHABILITATION AGENCY
Stephen W Hammonds Foundation	10010 Broadwa Street Apt. 805	San Antonio	TX	78217	FUND RAISING ORGANIZATION, NON-FEE BASIS
Abrazo Adoption Associates	10010 San Pedro Street 540	San Antonio	TX	78216	ADOPTION SERVICES
Council Oaks Cmnty Options LLC	10026 Trout Ridge Drive	Converse	TX	78109	RESIDENTIAL CARE FOR THE HANDICAPPED
Ernest C Olivares Senior Commu	1003 Vera Cruz	San Antonio	TX	78207	SENIOR CITIZENS' CENTER OR ASSOCIATION
First Aid Remodeling	10034 Aztec Village	San Antonio	TX	78245	GERIATRIC SOCIAL SERVICE
Los Vcinos De Las Misiones Cdc	10040 Espada Road	San Antonio	TX	78214	INDIVIDUAL AND FAMILY SERVICES
Fun Factory Centers	1006 Hammond Avenue	San Antonio	TX	78210	CHILD DAY CARE SERVICES
Itty Bitty Babies	1007 Division Avenue	San Antonio	TX	78225	CHILD DAY CARE SERVICES
Martin-Glasscock Neighborhood	101 Saint Francis Avenue	San Antonio	TX	78204	INDIVIDUAL AND FAMILY SERVICES
Hdqtrs A Force Svcs Agcy Svffp	10100 Reunion Place Street 400	San Antonio	TX	78216	INDIVIDUAL AND FAMILY SERVICES
Playtime Inflatables	1011 East Southcross Boulevard	San Antonio	TX	78214	CHILD DAY CARE SERVICES
Ramirez Center	1011 Gillette Boulevard	San Antonio	TX	78224	SOCIAL SERVICES, NEC
Colonial Home Health Inc.	1014 Camaron Street	San Antonio	TX	78212	GERIATRIC SOCIAL SERVICE
San Antonio Field Office North	1015 Jackson Keller Road	San Antonio	TX	78213	VOCATIONAL REHABILITATION AGENCY
Mental Health Association	1017 North Main Avenue Ste. 300	San Antonio	TX	78212	REFERRAL SERVICE FOR PERSONAL AND SOCIAL PROBLEMS

0 (0 out of 20 Selected)

Agencies$

***TIP** Is it necessary to clean up addresses? No, because an address locator knows how to read addresses. There is no need to remove RM or STE for room and suite, respectively. These pieces of information are common parts of many addresses. There is no need to standardize Avenue or AVE or AV.—the software can read all versions of this common street identifier. Unless something is obviously incorrect, there's not much need to clean up the addresses.*

5. **Close the address table.**

Sign in to ArcGIS Online via ArcMap

1. **On the far left of the menu bar in ArcMap, click the File tab.**

2. **Click Sign In.**

3. **Input your ArcGIS Trial Subscription ID (the one you just created under "Set up ArcGIS Trial").** If you forgot your password, here's a hint: it must be at least nine characters and contain both numbers and letters. You do not get confirmation that you are logged in, but you can check by reselecting the File drop-down arrow. If you successfully logged in, it will say Sign Out - <username>.

Geocode addresses

NOTE: *You must have an internet connection to complete this exercise and Internet Explorer installed on your computer.*

1. **In the table of contents, right-click the Agencies$ file name, and click Geocode Addresses.**

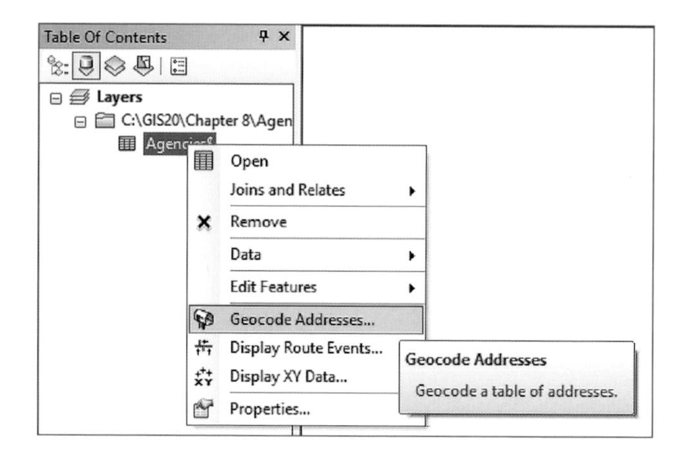

2. **In the "Choose an Address Locator to use" window, click Esri World Geocoder, and click OK.**

Untitled - ArcMap
File Edit View Bookmarks Insert Selection Geoprocessing Customize Windows Help
Table Of Contents
Layers
C:\GIS20\Chapter 8\Agen
Agencies$
Choose an Address Locator to use...
Name Description
* MGRS (Military Grid Reference System) MGRS Coordinates
Esri World Geocoder
Add...
OK
Cancel
-485.632 254.789 Unknown Units

NOTE: *If you do not see this address locator, it means you are using an older version of the software, and you should install the trial software version available with this book.*

In the tool's Address field, Address is auto-filled because the software looks for this header to fill in. If your addresses are in a column called something else, you must manually input which of your columns the proper address field is. There's a scroll bar on the right. If you scroll down, you'll notice City, State, and Zip are also filled in.

The next step is a little tricky. The "Output shapefile or feature class" field indicates where you want to save what will be the new geocoded file.

3. **Click the Browse button, and navigate to your save folder.** *Before* **you click save, you must change "Save as type" to Shapefile, instead of File and Personal Geodatabase feature classes. Click Save, and click OK.** Geocoding will begin (if you're lucky and have done everything right).

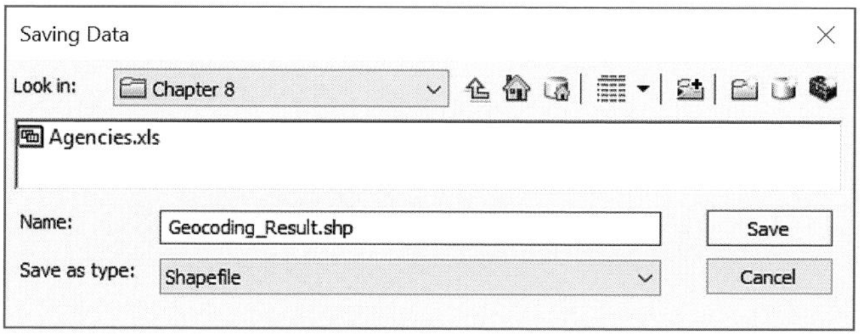

Saving Data
Look in: Chapter 8
Agencies.xls
Name: Geocoding_Result.shp
Save as type: Shapefile
Save
Cancel

You may encounter a Create Feature Class error when trying to geocode. It usually means you do not have enough credits to geocode. Even though the software lets you log in and seems as if it will geocode, it won't. It's all about those credits. If you are 100 percent sure you have enough credits, other possible issues are that you must have the latest version of Internet Explorer installed on your computer, and you must also be connected to the internet. Also realize that this geocoding won't work if you did not use the specific address locator Esri World Geocoder. You must upgrade your software to the latest version—or create your own address locator, as explained later in this chapter.

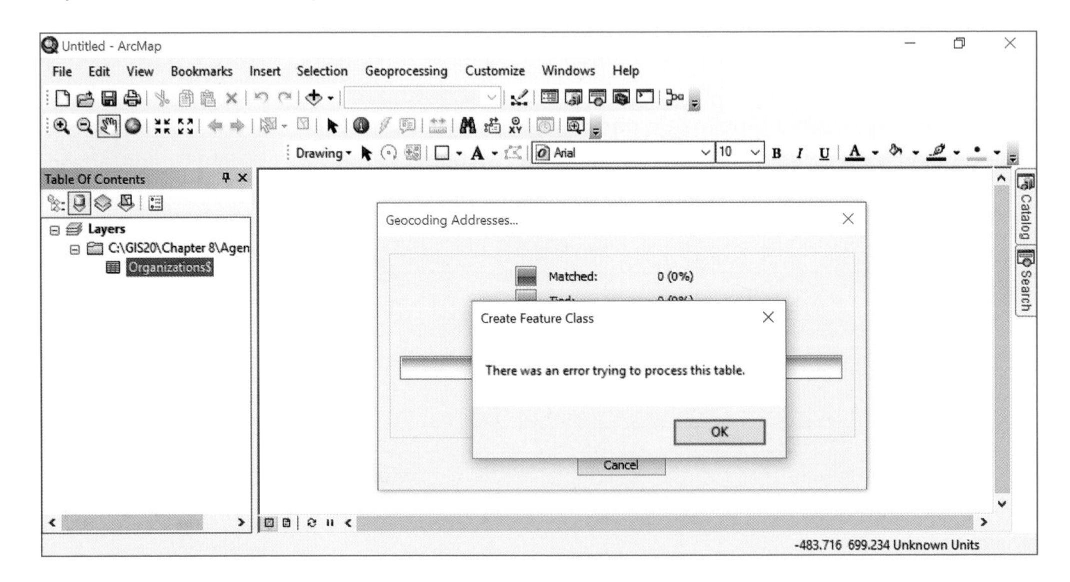

4. **Once the geocoding is complete, notice the statistics box, and see that 20 addresses were matched successfully.**

5. **Click OK, and then have a look at the map to see your beautiful dots.**

Change symbols

Hundreds of symbols are available in ArcMap. So now you can change the geocoded symbols to another shape.

1. **Right-click the geocoded file in the table of contents, click Properties, and then click the Symbology tab.**

2. **Click the symbol swatch, and notice all the symbols on the left side. Circle 3** Circle 3 **is a good choice, or whatever you like. Click OK twice.**

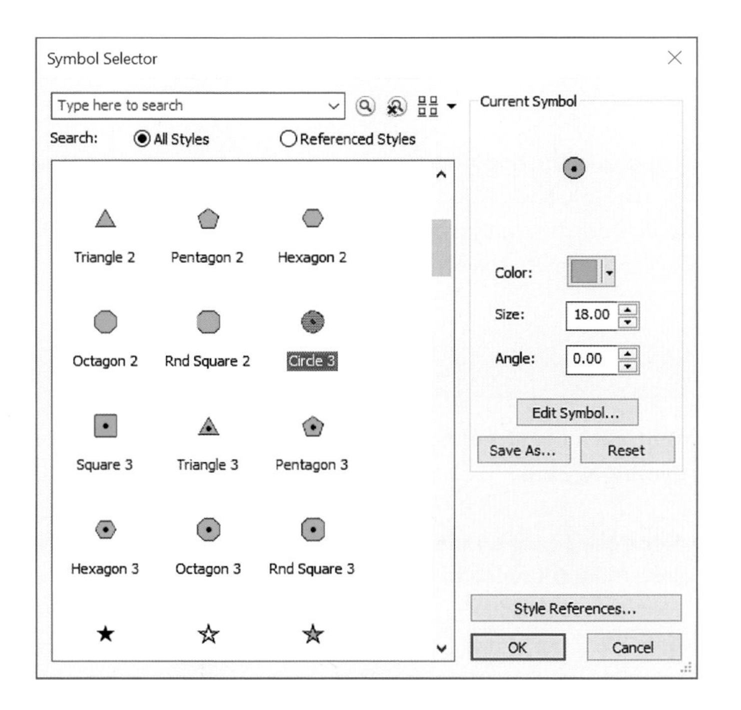

WANT OTHER SYMBOLS?

Click the Style References button, select Crime Analysis, and click OK. Notice that many new interesting symbols have been added to the Symbol Selector. Feel free to look at other symbol palettes.

Add other shapefiles

Geocoded maps benefit from context so your audience knows what they are looking at and where it is in relationship to everything else. Typically, streets and water go with geocoded addresses.

The social services for this exercise are in San Antonio, which is in Bexar County, Texas. To download a street network and water layer yourself, follow the steps in chapter 1, "Download county, city, and state shapefiles."

1. **Download the following layers:**
   - **All Lines—this is the street network (the downloaded name of the file though is tl_2016_48029_edges).**
   - **"Water"—select the Area Hydrography type (the downloaded file name is tl_2016_48029_areawater).**

NOTE: *You must unzip the files before you can use them in ArcMap.*

If you prefer to just grab the files, they are provided at esri.com/GIS20-3, in \GIS20\chapter 8, with the preceding names.

2. **In ArcMap, click Add Data, and add the two files to your map. You can add both at one time by pressing and holding the Ctrl key and clicking both, and then clicking the Add button. When you get the Geographic Coordinate Systems Warning, click Close to continue with the add.** Don't worry—ArcMap will conform to the projection that the geocodes are in.

Symbolize other layers

1. **In the table of contents, right-click the street network (also known as the edges file), click Properties, and click the Symbology tab.**

2. **Click the symbol swatch. From the styles on the left, select Residential Street, and change the color to a superlight gray. Click OK twice to close the open windows.**

3. **In the table of contents, right-click the water layer. Change the fill and outline color to a dark blue.** The outline around the water should be the same color as the water itself.

4. **In the table of contents, click the List By Drawing Order button.**

5. **In the table of contents, pull the water layer into first position.**

Save the project

1. **On the File menu, click Save As, and navigate to your save folder. Name the project** agencies.mxd**.** If you're finished learning about geocoding, you can stop and close ArcMap. If you'd like to learn to create your own address locator, or a little more about the ins and outs of geocoding, by all means, please continue reading.

Geocoding by creating your own address locator (labor required)

Create your own address locator

You are going to need a special tool for this task—ArcCatalog.

1. **Open ArcCatalog by clicking the Windows tab on the menu bar and selecting Catalog. You can also click the Catalog button .**

2. **Click the plus sign next to Folder Connections** Folder Connections **. You should see your save folder. Right-click on the file path (for example, C:\GIS20), click New, and then click Address Locator.**

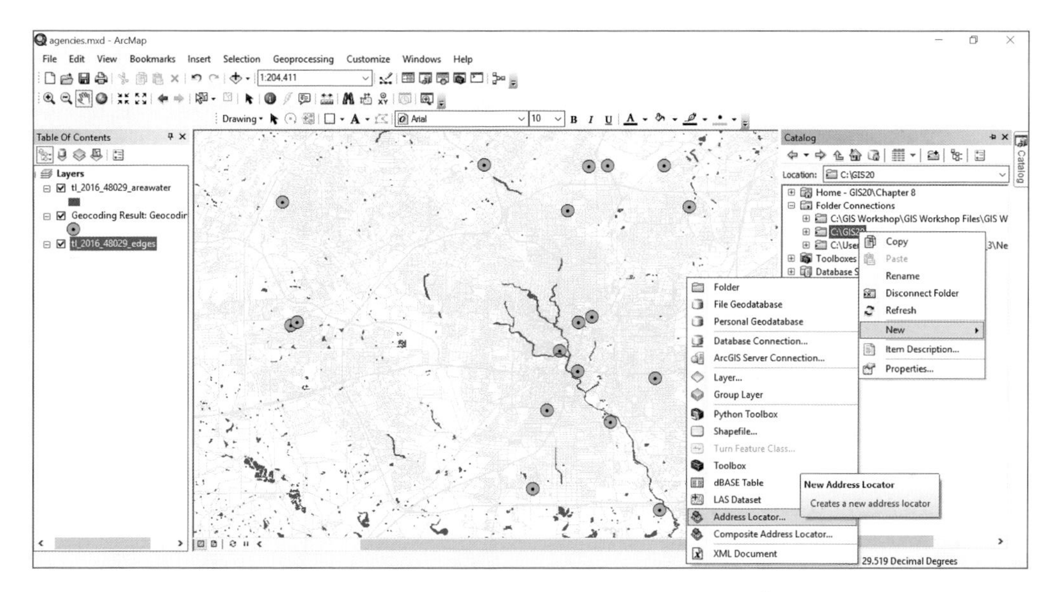

3. **In the first field, Address Locator Style, click the Browse button , and select US Address - Dual Ranges. Click OK.** A warning message will appear at the top in yellow, but you'll fix that in step 5.

4. **In the Reference Data field, click the drop-down arrow, and select the tl_2016_48029_ edges file.** If you didn't add this file to ArcMap earlier, you can download it from \GIS20\chapter 8, and add it to ArcMap. You'll see another red circle indicating an error. Don't worry! Keep going.

5. **First, on the Field Map box, pull down the bottom of the box to allow viewing of the full Field Map list. Under the Alias Name column, you must scroll down to the Street Name field in the Field Map, and change the None field to** FullName**.** Yes, there is already a Full Street Name field with the FullName field already filled in, but you're looking for the Street Name field three rows down from that one.

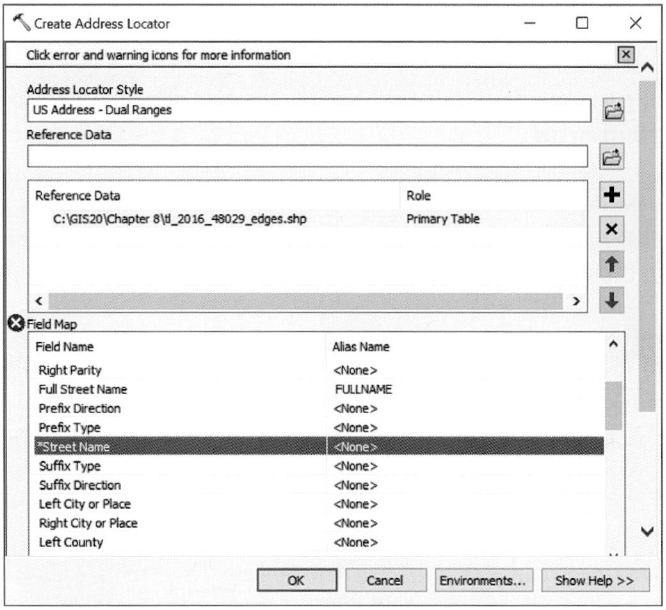

6. **Adjust all the following fields that have an asterisk (*):**
   - ***From Left to** LFromAdd
   - ***To Left to** LToAdd
   - ***From Right to** RFromAdd
   - ***To Right to** RToAdd
   - ***Street Name to** FullName

   The error messages should disappear once all the aliases are filled in.

7. **Scroll down to the last field, Output Address Locator, and click the Browse button to navigate to your save folder on your C drive. Give it the name** This is my Address Locator**, so it will be easy to spot when you need it next.** Your address locator should look like the one shown in the figure, although your file paths for reference data and output address locator fields will likely be different.

8. **Click OK.**

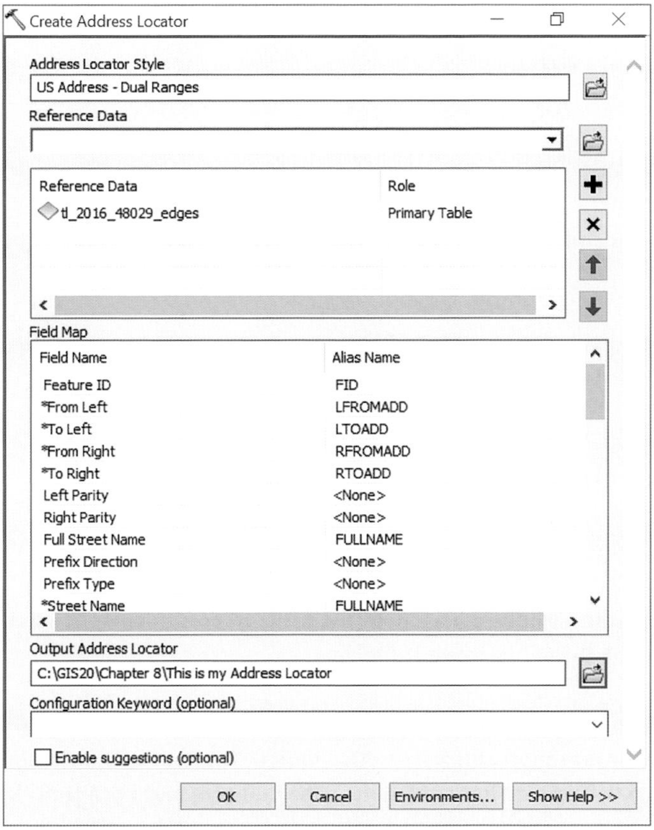

The creation of the address locator can take up to a minute. Once it's complete, you'll see a green check mark in the lower-right corner indicating the process has finished. If you expand the folder contents, you should be able to see a little red icon, which is your new address locator.

9. **Close ArcCatalog.**

Open new map

1. **Start a fresh mapping session by going to File > New > Blank Map, and click OK. Click No when asked if you want to save changes.**

Geocode using your address locator

1. **Do all the steps noted for geocoding, except for setting up the ArcGIS Trial and signing in to ArcGIS Online.**

2. **For Geocode Addresses, when you choose the address locator, click Add, navigate to where you saved it, and select "This is my Address Locator" for the address locator.**

Manually geocode unmatched addresses

You may have noticed that one of the addresses did not geocode.

1. **To figure out which address didn't have a match, in the table of contents right-click on the geocoded shapefile, and click Open Attribute Table.**

2. **In the attribute table, look in the Status column and note the *U* for one address, for unmatched, versus the *M* for matched addresses. Also look in the score column and see that it has a match score of 0, and also in the Match_addr column, and note that there is no address to which it's matched.**

3. **To determine the issue, scroll to the far right on the unmatched line item, and look at the address.** It is 10010 Broadwa Street Apt. 805—the *y* is missing from Broadway; therefore, the locator was not able to properly place the address.

4. **Close the attribute table.**

5. **In the table of contents, right-click on the geocoded file, and click Data > Review/Rematch Addresses. The Interactive Rematch box will pop up. Drag the lower-right corner of the box to make it larger so you can see what's going on.**

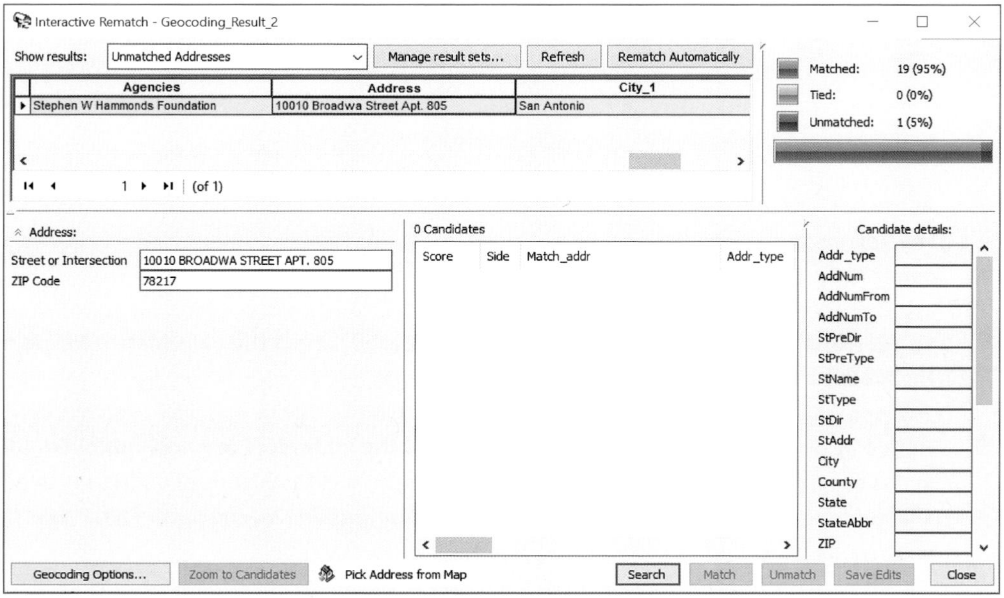

At this point, you might be a little confused, and it might be unclear what to do. The first record you see, the one highlighted in bright blue (this color is technically called "cyan"), is the unmatched address. But you won't be able to see the address unless you scroll to the far right.

6. **Scroll over so you can see the actual unmatched address (10010 Broadwa Street Apt. 805).**

Fix addresses

1. **To widen the Address section so you can see the address clearly, use your mouse to drag the light-gray line to the right of the section.**

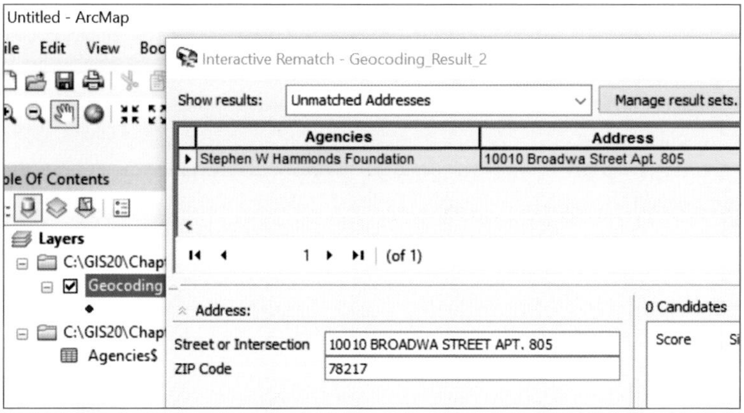

2. **In the field Street or Intersection (under the word "Address"), type over the given address.** In this example, the locator is confused because the street name is misspelled.

3. **In the Street or Intersection box, add** Y **to the end of BROADWA.**

4. **At the bottom of the screen, click Search.** The candidates listed will refresh with new matches. The first candidate has the highest score and is usually the best option—in this case, it is the correct address.

5. **To put a dot on the map where the address is located, select the candidate address, and click the Match button.**

6. **Notice in the upper-right corner, 100 percent of the addresses are now matched. Click Close.** In this case, you had only one address to fix, but if you had thousands of addresses to geocode and hundreds that had to be corrected, you could be working on the problem for days if you had to do it manually.

7. **Close ArcMap, and take a nap.**

HOW IMPORTANT IS IT TO GEOCODE 100 PERCENT OF ADDRESSES?

Suppose you had 5,000 addresses to geocode using a noncomposite address locator. That task would likely leave at least 10 percent, or 500 addresses, that would need to be corrected. It is an entirely subjective decision whether you should put in the effort to correct these 500 addresses, or whether you instead add a footnote that lets your readers know that only 90 percent of the addresses were geocoded. As with everything, the purpose of your map will determine whether correcting 100 percent of the addresses will make a difference or not.

GOOD TO KNOW

If you have several unmatched addresses after your geocoding session, you may prefer to export the attribute table of the geocoded shapefile and edit it in Microsoft Excel. To do so, sort the Status column descending with the *U* values at the top. The *U* values are unmatched addresses (*M* means matched). Highlight the records you want to export, click the Table Options button in the upper-left corner, and click Export.

Export all the records as a dBase table (.dbf). Now, go find someone to clean up all the addresses. Then re-geocode them.

CHAPTER 9

KEY CONCEPTS
mapping by category
understanding advanced symbology
grouping

Creating a categorical map

Categorical mapping is similar to thematic mapping in that color shading is used to indicate values; however, values in a categorical map represent categories instead of numbers. This technique can be used with polygons—for example, to map land-use zoning categories (residential, commercial, and industrial). It can also be used with point data, such as a geocoded file, to map such things as crime (burglaries and assaults). Categorical mapping also works with line data to map different types of streets (residential, major arterial, and highways).

In this exercise, you will map agencies by type of service, learn when categorical mapping is useful, and learn about different symbols.

[*You will need to download all chapter data files from esri.com/GIS20-3.*

Review attributes

1. **Open ArcMap. On the File menu, click Open, and then navigate to your save folder. Select agencies.mxd.**

2. **In the table of contents, right-click the geocoded file name, and click Open Attribute Table.**

3. **Scroll far right, and review the Type column.** This column indicates the type of service each organization provides. Several types of services are provided. This type of information (names of things) lends itself to categorical mapping.

4. **Close the attribute table.**

Create a categorical map

1. **In the table of contents, right-click the new layer named Geocoding_Result, click Properties, and click the Symbology tab.**

2. **From the Show list on the left, click Categories, and then click Unique values. In the Value field, select the column of data to map, which is Type_1 at the bottom of the list.** The symbol can be changed to anything you like. For this exercise, a smooth circle is preferable to the default dot.

3. **To change the symbol, double-click the dot next to the check box. In the Symbol Selector, select Circle 1 (you may need to scroll down, depending on what symbol palettes were selected in the last chapter), and change the size from 18.00 to 9. Click OK once.**

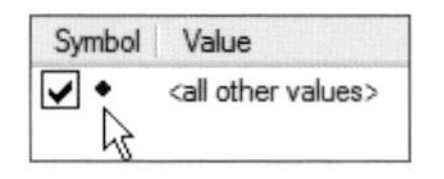

4. **Click the Add All Values button, and click OK to see colored dots displaying data.** The map will look like someone threw confetti (colorful dots) all over it. With this many categories and colors, the map is hard to read clearly.

5. **Select only a few types of services to display.**

Display only certain types of agencies

1. **In the table of contents, right-click the geocoded file name, click Properties, and click the Symbology tab.**

2. **Click the Remove All button to get rid of all the selections.**

3. **Click the Add Values button.**

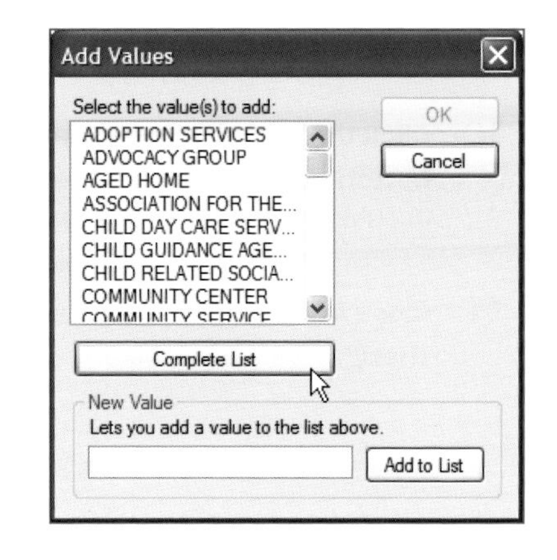

4. **Select the following three types of services: Aged Home, Geriatric Social Service, and Senior Citizens' Centers. Press and hold the Ctrl key to select more than one item at a time, or just select items individually and keep adding. Click OK.**

5. **Clear the <all other values> check box.**

6. **Double-click each dot, and change the dot color to something that will stand out. Click OK.**

Group several categories into one category

You can assign multiple categories to one encompassing category such as Senior Services.

1. **In the table of contents, right-click the shapefile name, click Properties, and click the Symbology tab.**

2. **Highlight the three categories by pressing and holding the Ctrl key as you click each category name.**

3. **Right-click within the dark-blue highlighted area, and click Group Values.** All three categories will now be grouped into one category (the name at this point doesn't matter).

4. **Click OK.**

5. **In the table of contents, click once on the grouped layer to activate the text. Rename this grouping** Senior Services**.**

6. **(Optional) To label some of these dots, use the Label tool, which can be found on the Draw toolbar. Access the Draw toolbar by going to the Customize menu, clicking Toolbars, and clicking Draw.**

7. **After the tool is activated, expand the capital A. Click the Label tool to activate it, and then click a dot on the map to label it.** Callout boxes around labels may be helpful.

8. **(Optional) To apply callout boxes, using the default pointer, select all labels (press and hold the Ctrl key to select more than one).**

9. **After the labels are selected, right-click any of them, and click Properties. Click the Change Symbol button, and then click the Edit Symbol button** Edit Symbol... **.**

10. **Click the Advanced Text tab, and select the Text Background check box. Click the Properties button underneath it. Select the second balloon callout, and click OK four times.** You may need to reposition the labels to make the anchor visible.

11. **Close ArcMap, or to proceed to the next exercise, click File > New > Blank Map, and click OK. There's no need to save.**

CHAPTER 10

KEY CONCEPTS

understanding latitude-longitude
using GPS units
creating new shapefiles

GPS point mapping

In addition to mapping a table of addresses (see chapter 8), you can also map a table of latitude and longitude (x,y) coordinates collected in the field using a Global Positioning System (GPS) unit. Latitude and longitude points are frequently used in GIS—for example, when a particular place doesn't have an address but you still need to display it as a dot on the map.

Point mapping is often used in environmental research to map such things as water quality samples, invasive plant species, or more recently, animal tagging (sharks and monkeys). In this exercise, you will learn to place latitude and longitude points on a map. You may never need to do this again—then again, maybe you will.

You will need to download all chapter data files from esri.com/GIS20-3.

Add latitude-longitude points

1. **First download the chapter 10 data files from the book resource web page. Then click Add Data , and double-click XYData.xls and add the desired worksheet, XYs$.**

2. **In the table of contents, right-click the file, and click Open. Have a quick look at the data table to confirm you have one X (longitude) column and one Y (latitude) column (as well as a couple of other columns). Close the data table.**

Place points on your map

1. **In the table of contents, right-click XYs$, and click Display XY Data.** The X and Y fields automatically fill in on the basis of column headers of X and Y. There is no Z field.

   The coordinate system must be defined.

2. **Click the Edit button. Expand the Geographic Coordinate System folder, expand the North America folder, and then click NAD 1983. Click OK twice.**

3. **When you get a message stating "Table does not have Object-ID field," click OK.** It will open with no problems.

   TIP *You selected NAD 83 (for more on projections, see chapter 3). NAD 83 is a common choice. If it doesn't work for your data, check your GPS to determine the geographic coordinate system that was used.*

Create a new shapefile of XY Events

Right now, the XY layer is not saved; it is only a temporary file.

1. **To create a new, permanent shapefile of these x,y coordinates, in the table of contents, right-click the XYs$ Events layer. Click Data > Export Data, and save the new shapefile.** You may need to reproject the shapefile to work with other layers.

2. **Close ArcMap, or to proceed to the next exercise, click File > New > Blank Map, and click OK. There's no need to save.**

GETTING LATITUDE-LONGITUDE POINTS OUT OF YOUR GPS UNIT AND INTO MICROSOFT EXCEL

Every GPS unit is different so it is difficult to give specific advice about how to get points from your GPS unit into Excel. One way is to use a tool in ArcGIS called GPX to Features, which allows you to quickly upload these points to ArcMap. If your GPS does not use the GPX file format, you must export points out of your GPS unit (perhaps as a text, comma-separated value, or database file), open them in Excel, do some minor cleanup work, and save them.

1. Click points with your GPS. These are sometimes called *waypoints* or *points of interest* (POIs).
2. Connect the GPS to your computer with a USB cord.
3. Navigate to the folder where your clicked points are stored in your GPS.

Your GPS likely came with software that must be installed on your computer ahead of time. If it has been installed, you can navigate to this folder. Often each GPS unit uses a proprietary file format, so don't be surprised if you don't recognize the file extension of your point file.

You should be able to open the file in Notepad to make sure it is the correct file.

4. Try double-clicking the file, and when prompted for which software to use to open it, select Notepad. Save as a text file (.txt). Close the file.
5. Open your text file in Excel.

This process will vary depending on what version of Excel you are using. You may need to import the file into Excel, or you may simply be able to open the file.

6. Once your data is imported into Excel, identify your X (longitude or easting) and Y (latitude or northing) coordinate fields, and any corresponding attribute fields. If needed, add or modify the header row to give the columns easy-to-understand titles. Longitude should be titled *X*, and latitude should be titled *Y*.
7. Save as an Excel file, and name the worksheet.

Now the file is ready to open in ArcMap.

THE GEOTAGGED PHOTOS TO POINTS TOOL

The GeoTagged Photos to Points tool (in ArcToolbox > Data Management Tools > Photos) reads the XY data from photo files (.jpg and .tif only) taken with a digital camera or smartphone. Assuming the camera has the technology to grab the latitude and longitude of the place where the photo was taken, you can quickly and easily upload these points. And the cool part is you can then click on the points in ArcMap to see the photos.

CHAPTER 11

KEY CONCEPTS
changing boundaries
making custom selections
creating shapefiles from scratch

Editing

Sometimes, it's necessary to change the physical boundary of an existing polygon. For example, if your agency uses a target area boundary to deliver services and wants to extend the service area, you would have to edit the boundary. Target area boundaries might include such things as school districts, voting wards, and neighborhoods.

I bet you thought editing maps was only for those with a PhD in mapmaking. Well, you'd be wrong. In this exercise, you will learn common editing tasks such as changing a boundary outline, merging polygons, creating shapefiles out of selected polygons, and appending shapefiles to each other.

You will need to download all chapter data files from esri.com/GIS20-3.

Open a shapefile, and turn on the Editor toolbar

1. **Click Add Data to add States.shp.**

2. **Use the Zoom In tool to zoom in closer to Oregon (or whichever state you want) so you can see the outline well.**

3. **Click the Editor Toolbar button .** The Editor toolbar will become active. The Editor Toolbar button should already be visible because it is a part of the default ArcMap interface (if it's not, go to Customize > Toolbars > Editor). If you have an older version of the software, go to the View menu, click Toolbars, and click Editor.

Edit the state outline

1. **Click the Editor button Editor, and click Start Editing.** Not only the polygon boundaries but also the attribute table become editable. A little arrow appears and serves as a pointer. This pointer is the Edit tool.

2. **With this arrow, double-click the polygon to edit, in this case Oregon, and notice how several little dots (called *vertices*) appear in green. Zoom in very close so you can see them.** By moving these dots, you can reshape the boundary of the state.

3. **Use the Zoom In tool to zoom in close (or the mouse scroll wheel) to see the dots clearly. You may need to reactivate the Edit tool after using the Zoom In tool.**

4. **Click any one of the dots, and drag it to a different position to begin reshaping. Try this with a few more vertices. When you are finished, click anywhere outside the state boundary to complete the edits.**

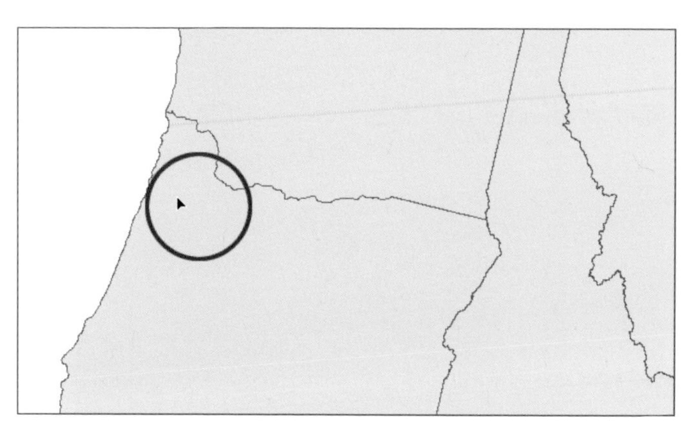

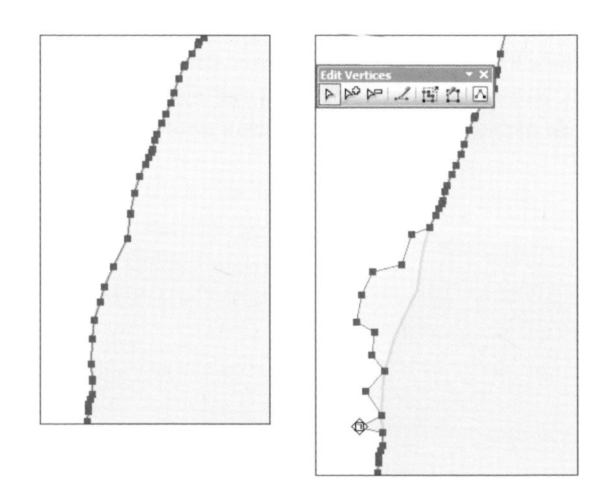

Move polygons and cut

Besides moving vertices, you can move an entire polygon at once.

1. **Zoom out to see multiple states. Click the Edit tool again, and click Oregon. Notice the state becomes highlighted in bright blue.** This highlighting means the entire polygon is selected, and you can move that polygon.

2. **Move Oregon to the left so it doesn't touch the US anymore.** The Editor toolbar contains additional tools. One you may find handy is the Cut Polygons tool.

3. **With Oregon still highlighted and not touching the US, click the Cut Polygons tool to activate it.**

4. **Click the left side of Oregon, and draw a line across Oregon toward the right. On the right side, double-click the Oregon border. Notice, the state has been cut in two. Click the Edit tool again, and click on the lower portion of Oregon. Drag that portion down, to see the split.**

5. **Click the Editor button again, and click Stop Editing. You do not need to save your changes.**

Create new shapefiles out of existing ones

1. **Click the Select Features by Rectangle tool .**

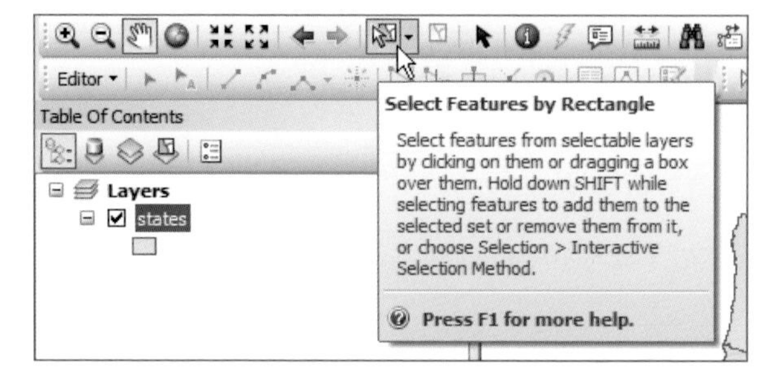

2. **Press and hold the Shift key, and click on each state to select it (California, Nevada, Utah, and Arizona).** These states will be highlighted in bright blue.

3. **In the table of contents, right-click States.shp.**

4. **Click Data, and then click Export Data. Click the Browse button to navigate to your save folder, and type a name. Unless your file path already points to your save folder, you can simply type over "export-import.shp," and give the file a new name. Ensure that "Save as type" is Shapefile. Click Save, and click OK.**

5. **Click Yes to add the data as a layer.** A new layer is added to ArcMap.

6. **To remove the bright-blue outline, on the Selection menu on the menu bar, click Clear Selected Features. Or you can click the Clear Selected Features button on the Tools toolbar. Notice if you clear the check box next to States in the table of contents, you can see just the new shapefile.**

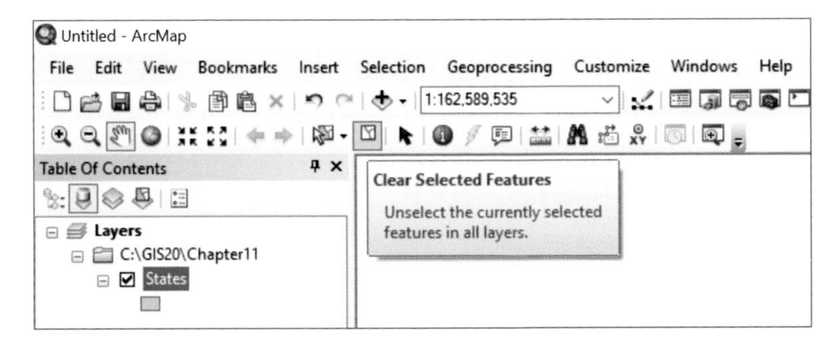

7. **In the table of contents, right-click the new shapefile and remove it, leaving only the States shapefile. Turn the States layer on, and you can see the new shapefile.**

Create new features on an existing shapefile

Pretend someone has said you can create a brand-new state. Of course, the first step toward statehood would be to draw a boundary for the new state.

1. **Click the Editor button** Editor **, and then click Start Editing to make the shapefile editable.**

2. **At the end of the Editor toolbar, click the Create Features button .**

3. **In the first box, click States, and select Polygon as the construction tool.**

Create Features
<Search>
states
states
Construction Tools
Polygon
Rectangle
Circle

4. **Move the Create Features box slightly out of the way so you can see the map.**

5. **Begin drawing a new state next to California. Double-click the polygon when you are finished, and you will see a new shape.** The new state boundary may not be a perfect fit with the California boundary. This is a great time to explore some of the other editing tools.

6. **Click the Edit Vertices button . Then click the Add Vertex button (it's the one with a plus sign) to add additional vertices along the edge that borders California to modify it and make it a better fit.**

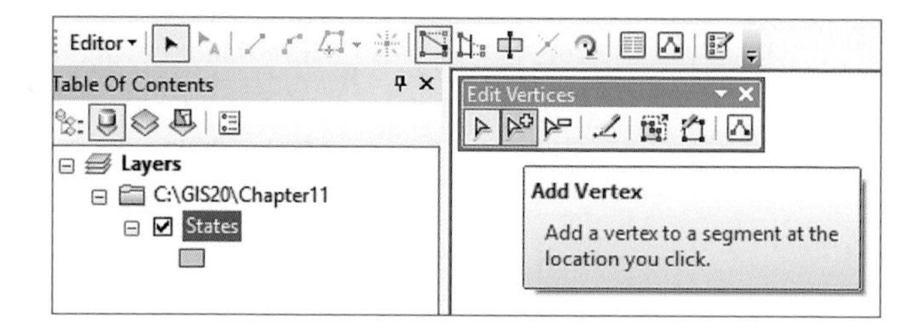

TIP *The attribute table for states is also now editable, and you can associate information with this new feature, including giving it a name.*

7. **Stop editing. You do not need to save your edits.**

Create a new shapefile

Occasionally, it may be necessary to create a new shapefile out of thin air.

1. **Click the ArcToolbox button to open ArcToolbox.** The toolbox may take up to a full minute to open—that's normal.

2. **Expand the Data Management Tools toolbox, and then expand the Feature Class toolset.**

3. **Double-click the Create Feature Class tool.**

4. **In the Feature Class Location field, click the Browse button . Navigate to the folder where you will save what will be the newly created shapefile.** Getting to the folder can get tricky.

5. **Click the drop-down arrow on the right of Look In and under Folder Connections, selecting the folder where you want to save your file.** If your folder does not have any such folders and you simply click on the GIS20 folder, you will get a pop-up window, which looks as if you are meant to type a name of the file.

   But once you type the name and click Add, you get the error "Object Not Found." This error is because you have clicked *into* the folder, and not merely *on* the folder.

6. **Click Cancel on the error message, and click the Browse button again. Use the drop-down arrow to click once on the folder name, and then click Add.** You might find it helpful to use the Up One Level tool to get to a spot where you can simply click the folder name, without going into the actual folder. Assuming you could get through this step, you can proceed to the next one.

7. **In the Feature Class Name field, type** newsites **(avoid spaces in file names).**

8. **In the Geometry Type field, select Point.** The layer you'll create in a few steps will be composed of points, not polygons.

9. **Do not select anything for the Template Feature Class field.**

10. **Next to the Coordinate Systems field, click the Coordinate System button .**

11. **Expand the Projected Coordinate Systems folder.**

12. **Select Continental > North America > North America Albers Equal Area Conic. Click OK twice.** The file newsites.shp should be added to the table of contents. This shapefile is empty. This process will take a minute.

13. **You can close ArcToolbox. You won't be using it again in this exercise.**

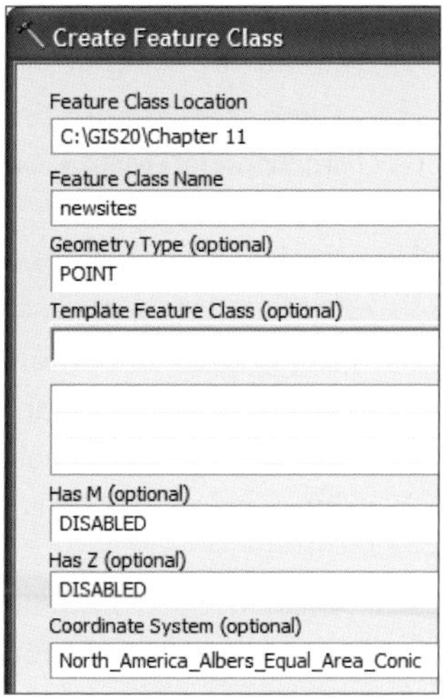

Create Feature Class
Feature Class Location
C:\GIS20\Chapter 11
Feature Class Name
newsites
Geometry Type (optional)
POINT
Template Feature Class (optional)
Has M (optional)
DISABLED
Has Z (optional)
DISABLED
Coordinate System (optional)
North_America_Albers_Equal_Area_Conic

Construct a new shapefile

1. **Click the Editor Toolbar button, and click Start Editing. You must click which file to edit so click newsites.**

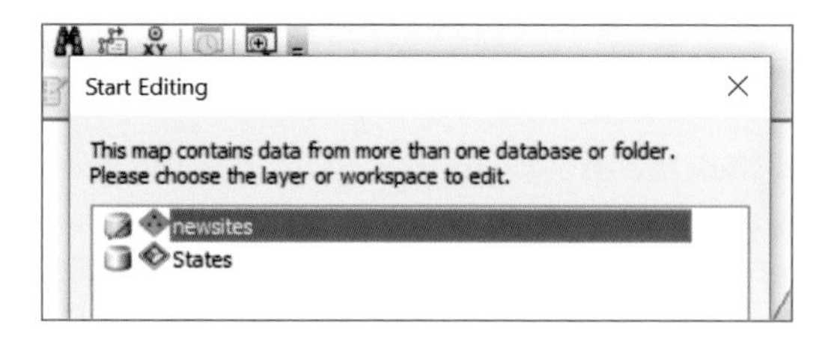

2. **Click the Create Features tool, and notice a new Create Features box is added on the right.** You are going to add new points to the blank shapefile.

3. **In the Create Features box, notice newsites and States. Click newsites, and notice what tools appear under Construction Tools at the bottom. In the Create Features box, click the newsites file name, and select the Point tool. (If you wanted to create a polygon-based shapefile, you must begin the editing with a polygon file, such as States.)**

4. **If the States map is not already zoomed out, zoom out on the map so you can see the whole US.** The mouse scroll wheel works great to zoom out.

5. **Click anywhere in Texas to place a new point. Now place two other points, one in California and one in Florida.** Close the Create Features box to get it out of the way.

   You are not finished yet. Next, you will add names in the attribute table for these new sites.

6. **In the table of contents, right-click the newsites layer name, and then click Open Attribute Table.** No useful options exist for inputting names for each of these new sites. You can create a column for site names. To add a column to the attribute table, the Editor button must be turned off.

7. **Close the attribute table. Click the Editor button, and click Stop Editing. Click Yes when prompted to save your edits.** Now you can add a column to the attribute table.

8. **In the table of contents, right-click the newsites layer name, and then click Open Attribute Table.**

9. On the Table Options menu (the first button), click Add Field.

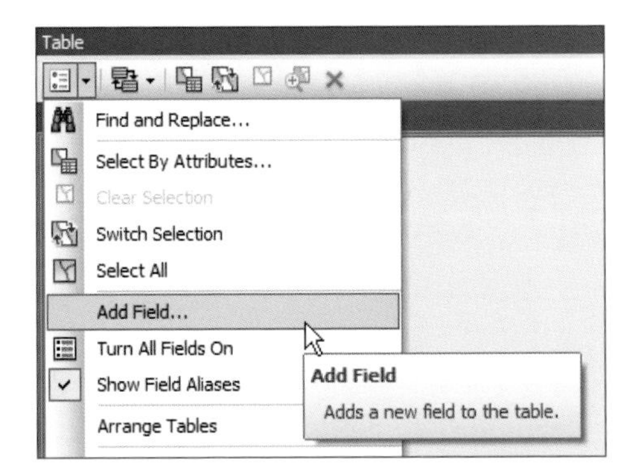

10. For Name, type Name**, as this will be the name for the new sites. For Type, select Text, as this column will contain only text.**

11. Leave the length as 50, and click OK. Notice a new column has been added. Close the attribute table.

12. Click the Editor button, and click Start Editing to make this layer editable.

13. On the Editor toolbar, click the Attributes button. Notice a box appears for attributes.

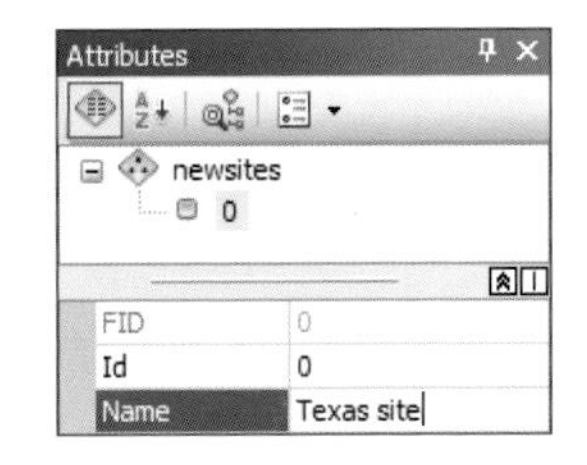

14. Click the first point in Texas, and type Texas site **in the Name field in the Attributes panel.**

15. Click the California and Florida sites, and do step 15 for those points.

16. When you are finished editing the attributes, click Editor > Save Edits > Stop Editing. Close the Attributes box.

17. Close ArcMap, and go get a pizza. Or to proceed to the next exercise, click File > New > Blank Map, and click OK. There's no need to save.

CHAPTER 12

KEY CONCEPTS

writing and erasing queries

creating new shapefiles on the basis of query results

Creating attribute queries

The ability to query data in an attribute table separates GIS software from graphic design software. Having the ability to query selections of data allows users to analyze communities and problems in more sophisticated ways, use the software to ask questions, and perform intelligent analysis on the basis of the answers.

Mapping is not just about mapping. It's about asking complex questions of your data, and getting a useful answer. It's making your data work for you. In this exercise, you'll learn to do the most important query in all of ArcGIS—attribute queries.

You will need to download all chapter data files from esri.com/GIS20-3.

Add shapefiles

1. **Click Add Data ✦ ▾ to add agejoined.shp, which was created in chapter 5. This file is also downloadable at esri.com/GIS20-3 in the chapter 12 folder.**

Write a query

1. **On the Selection menu, click Select By Attributes.**

2. **Enter the following query:** "Seniors" > 15**. Click OK.** Counties that meet this condition are highlighted with a bright-blue line.

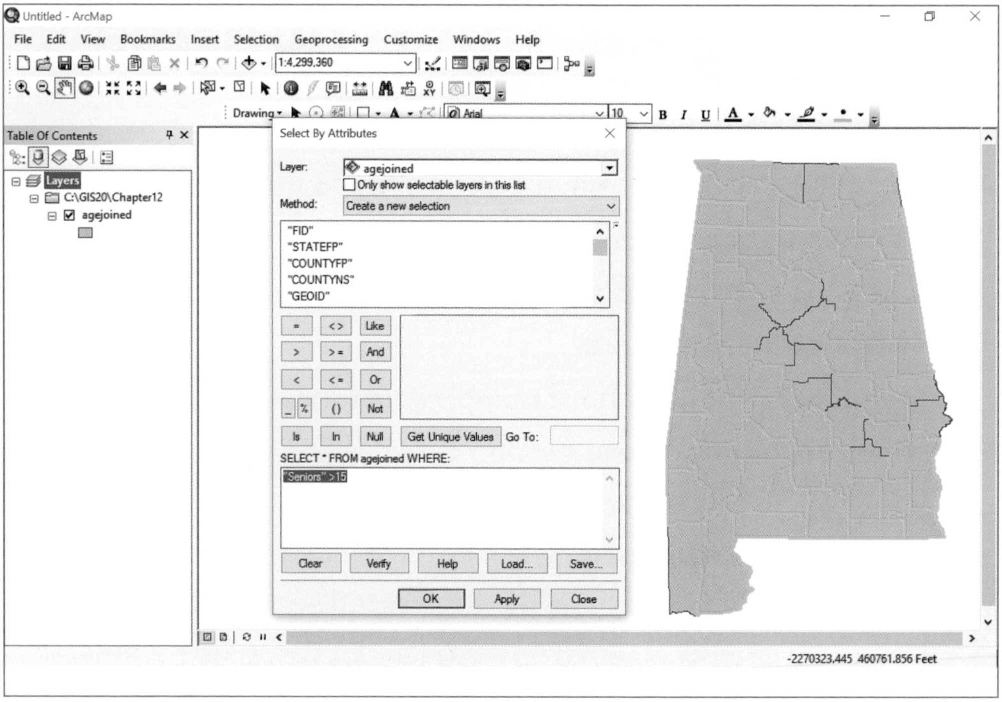

3. **Open the attribute table, and notice the corresponding rows are also highlighted. (See "Export selected records.") Close the attribute table.**

Create a new shapefile for selected counties

1. **In the table of contents, right-click the layer name, click Data, and then click Export Data.**

2. **Navigate to your save folder. Name the file** seniors**. Save as type should be Shapefile. This file will be used in the next exercise, in chapter 13, so note where it is saved. Click Save, and click OK.**

3. **When prompted, click Yes to add the shapefile to the table of contents.**

4. **In the table of contents, clear the check box next to agejoined to turn it off so you can clearly see what the new file contains.** This process of how to create a shapefile and export a selection of records is how to create a "target area" map on the basis of data and a query of the attributes.

EXPORT SELECTED RECORDS

It is easy to export a selection of records on the basis of query results. After the query is completed, do the following steps:

1. In the table of contents, right-click agejoined, and open the attribute table.
2. On the Table toolbar, click the Table Options tool arrow, and then click Export.
3. Save the file.

The file type should be .dbf. This generic database file type can be read by many software programs, including Microsoft Excel and Microsoft Access®.

Erase the query

1. **To erase the query, on the Selection menu, click Clear Selected Features. Alternatively, click the Clear Selected Features button.** When the blue highlighting is gone, the query is erased, and nothing is selected.

2. **Close ArcMap, or to proceed to the next exercise, click File > New > Blank Map, and click OK. There's no need to save.**

CHAPTER 13

KEY CONCEPTS
writing queries
understanding centroids
understanding intersect

Creating location queries

Location queries differ from attribute queries in that they do not involve selecting data. A location query is a geography query, not a data query. A location query involves selecting geographies within other geographies. This type of query works equally well with point, line, or polygon data to find locations.

This exercise is the yin to the attribute query's yang—it's a location query, the second most awesome of all the queries.

You will need to download all chapter data files from esri.com/GIS20-3.

Add shapefiles

1. **Click Add Data , and add your place shapefile downloaded in chapter 1, or use the Alabama place file named tl_2016_01_place.shp in chapter 13 in your save folder.**

2. **Next, add seniors.shp, which was created in chapter 12. (There's also a Seniors shapefile for Alabama you can use in the \GIS20\Chapter13 folder.)**

3. **In the table of contents, click the List By Drawing Order button , and then drag places into first position in the table of contents.**

Create a location query

The object is to create a query that will select cities within the boundaries of the counties in the Seniors shapefile.

1. **On the Selection menu, click Select By Location.**

2. **For the section underneath Target Layer(s), select the check box for your place file—if you're using the Alabama file, it's tl_2016_01_place.**

3. **For the Source layer, select the seniors shapefile.**

4. **For "Spatial selection method for target layer feature(s)," select "have their centroid in the source layer feature." Click OK.**

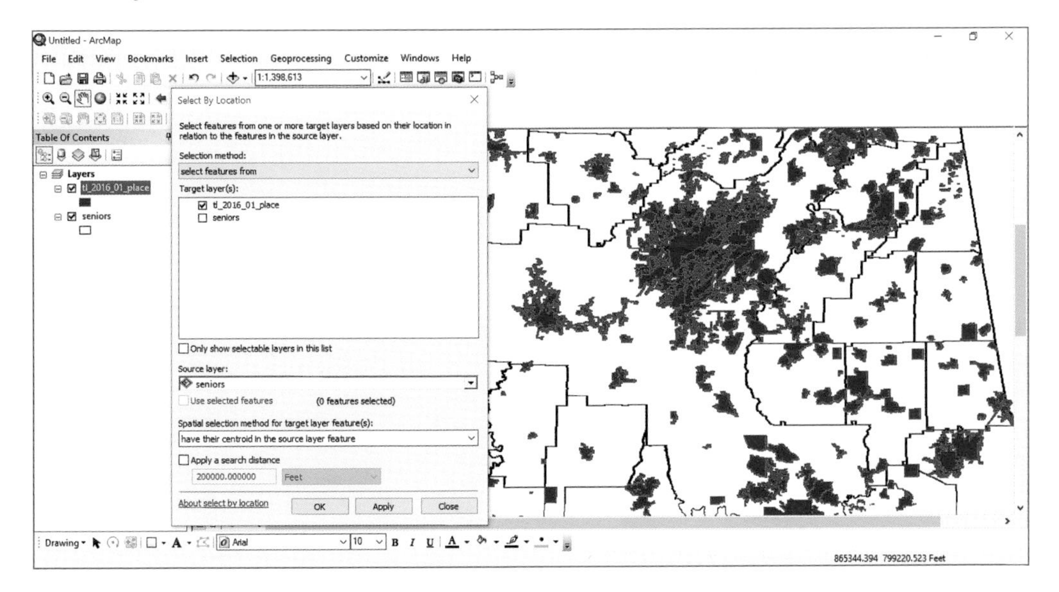

Only those places within senior counties are selected. All other cities outside the senior counties are ignored.

5. **In the table of contents, right-click the place layer name, and open the attribute table. Now you can see which cities were selected.**

NOTE: *For more on how to export only these selected records, see chapter 12.*

6. **Close ArcMap, or to proceed to the next exercise, click File > New > Blank Map, and click OK. There's no need to save.**

WHAT DOES "HAVE THEIR CENTROID IN" MEAN?

A centroid is the physical center point of geometry (in this case, a polygon). By using the "have their centroid in" method, the physical center of the place must fall within the county boundary, not merely touch or intersect it. This method is more conservative. You will never get polygons that are only slightly within the boundary because the center point of the polygon must fall within the target boundary.

Another popular tool to use for location queries is Intersect, which would include any place that intersects the county boundary.

CHAPTER 14

KEY CONCEPTS

buffering
merging
unioning
appending
clipping
dissolving

Using geoprocessing tools

Geoprocessing refers to various GIS operations that manipulate spatial data. These tools include buffer, merge, union, append, clip, and dissolve. The ability to perform complex spatial calculations sets GIS apart from traditional cartography. It is the muscle of GIS.

In this exercise, you will explore using GIS power tools for geoprocessing.

You will need to download all chapter data files from esri.com/GIS20-3.

Buffering

A buffer is a map item that represents a uniform distance around a feature (point, line, or polygon). When creating a buffer, the user selects the feature to buffer around, as well as the distance of the buffer.

Add a shapefile

1. **Click Add Data to add youthcenters.shp from the data files on the book resource web page. This file contains youth centers in Bexar County, Texas.**

Create a buffer

1. **To create a buffer of 1,000 feet around each of these youth centers, click the Geoprocessing tab on the menu bar, and then select Buffer.**

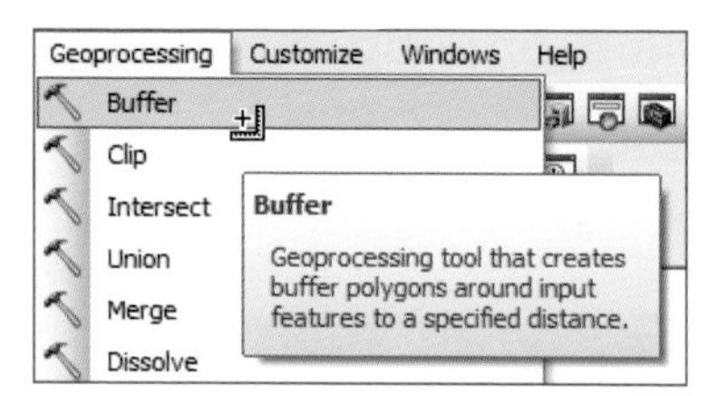

2. **For Input Features, select youthcenters.**

3. **For Output Feature Class, click the Browse button to navigate to your save folder.** A new buffer file will be created.

4. **Name the file** buffers**, and then click Save.**

5. **For the linear unit, type** 1000**. Ensure that the unit type is set to feet. Leave all other options set as they are, and then click OK.**

Add other files, and change the symbology

1. **From chapter 14 in your save folder, add street network tl_2016_48029_edges.shp.**

2. **Turn labels on for streets, change the street color to gray, and make the buffer hollow with a thick red border (all items covered in previous chapters).**

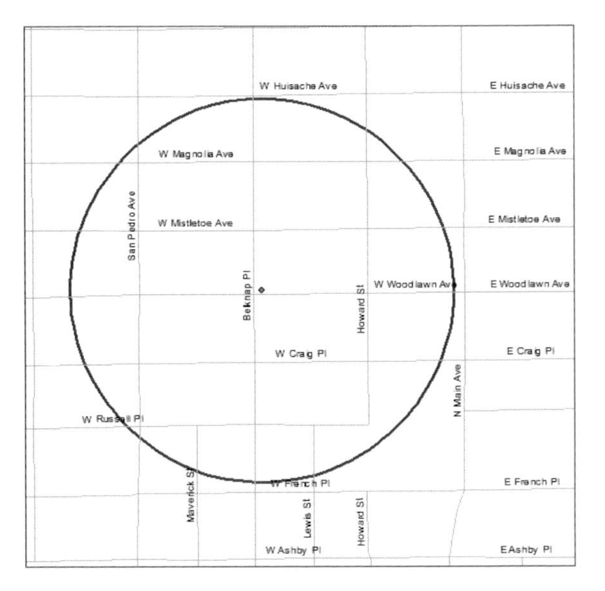

Another tool that may be useful for double-checking is the Measure tool.

3. **Click the Measure tool to enable it. Click the Choose Units button, click Distance, and ensure that Feet is checked.**

4. **Click a dot that represents a youth center, and draw a line to the edge of the buffer. Notice that the distance is 1,000 feet.**

5. **Close ArcMap. There's no need to save.**

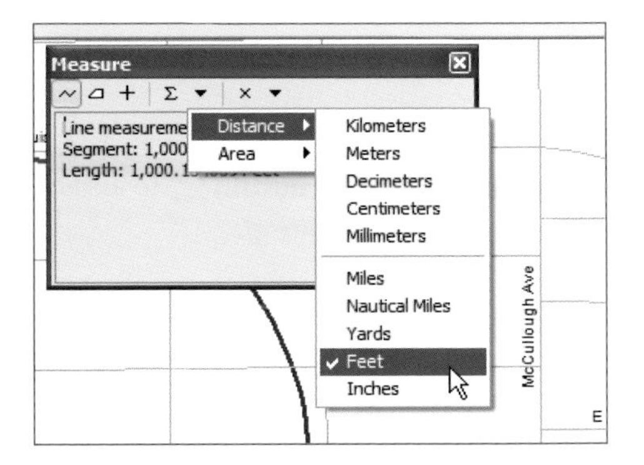

THE CHOOSE MEASUREMENT TYPE TOOL

Do you know spatial calculus? Me neither. That's okay because ArcMap does the work for us. The Choose Measurement Type tool gives you planar, geodesic, and a few other options. Planar uses 2D Cartesian mathematics to calculate distance and is used with projected coordinate systems, such as the one used in this chapter, state plane. The tool is designed to measure things on a flat surface. Geodesic measuring is a different type of measuring system that calculates the shortest distance between two points, considering that the ellipsoid shape of the earth is not perfectly round,.

The distance between two cities will not be the same if you (a) draw buffers using different underlying coordinate systems, and (b) measure using different measurement types. The geodesic system is thought to be the more accurate measuring system of the two. But in our example, there isn't much difference between planar and geodesic measurements because we created buffers for a small area and used a correct projection for the map.

Type **buffer analysis** in the Help menu, and you will find a lot of information about this topic. One quick tip: the larger the area, the more you should consider using a geographic coordinate system for buffering. Using a geographic coordinate system will minimize distance calculation errors.

Merging

Example of Merge

It is important to start with a fresh mapping session. If you did not close ArcMap from the buffer exercise, do so now. Alternatively, click File > New > Blank Map, and click OK. This action will open a new edit session.

Merge tool

1. **Use the Add Data button to add the states shapefile.** This file is of all the states within the US. You also used this file in chapter 11 while editing.

2. **Begin by making the shapefile editable. Click the Editor Toolbar button to activate the Editor toolbar. Click the Editor button** Editor ▾ **, and click Start Editing.**

3. **Press and hold the Shift key, and use the Edit tool to select multiple contiguous polygons, such as three different states (in this example, Oregon, Washington, and Idaho).** You can tell they are selected when all appear with bright-blue shading.

4. **Click Editor, and then click Merge.** Because all three states will now take on the attributes of only one state, you must select which state that will be.

5. **When you get the dialog box that says "Choose the feature with which other features will be merged," select one of the states.** Oregon was selected for this example.

6. **Click OK, and notice that all three states have merged into one.**

7. **Open the attribute table, right-click on the Name column heading, and click either Sort Ascending or Sort Descending.**

8. **Notice that now neither Idaho nor Washington are present.**

9. **Close the attribute table. Click the Clear Selected Features tool to clear the selection.**

Unioning

The Union tool differs from the Merge tool in that each record in the attribute table is maintained for each merged item; however, the visible boundary between the polygons disappears. Visually, a union looks like one big new polygon (same as a merge), but within the attribute table, a record is maintained for each polygon.

Union tool

1. **Note that the layer is still editable. Press and hold the Shift key, and with the Edit tool, select another three contiguous polygons.** In this example, Texas, New Mexico, and Oklahoma are selected.

2. **Click Editor, and select Union. When you get a dialog box that says, "Choose a template to create feature(s) with," click OK.**

3. **Notice the three states now appear as one; however, they are still three distinct states. Click anywhere outside the states to get rid of the selection (bright-blue highlight). Now click more or less in the area where Texas should be (or any one of the states you may have selected).**

4. **Notice how the original outline of that one state appears selected.** Visually, the outline appears to be one boundary, but the three states are all still separate line items in the attribute table.

5. **Open the attribute table to confirm you still have one line item for Texas, New Mexico, and Oklahoma. Close the attribute table. Click the Clear Selected Features tool to clear the selection.**

6. **Close ArcMap. There's no need to save.**

Appending

The Append tool combines multiple shapefiles into one shapefile.

It is important to start with a fresh edit session. If you did not close ArcMap from the union exercise, do so now. Alternatively, you can click File > New > Blank Map, and click OK. This will open a new edit session.

Append tool

1. **Download Pacific_NW.shp and Southwest.shp from the book resource web page, in chapter 14 of your save folder, if you haven't already. Click Add Data to add Pacific_NW.shp and Southwest.shp.** The Pacific_NW shapefile contains three states in the Pacific Northwest. The Southwest file contains southwestern states. The idea is to get these two separate files into one shapefile.

2. **Click the ArcToolbox button to open ArcToolbox. Expand the Data Management Tools toolbox, and then expand the General toolset. Double-click the Append tool.** Now you'll append the Southwest shapefile to the Pacific_NW shapefile.

3. **From the Input Datasets list, select Southwest. From the Target Dataset list, select Pacific_ NW shp.** This shapefile is where Southwest.shp will be appended.

4. **In the Schema Type field, leave the default Test. Click OK.**

Append

Input Datasets

Southwest

Target Dataset

Pacific_NW

Schema Type (optional)

TEST

Field Map (optional)

OK | Cancel | Environments... | Show Help >>

5. **Turn off the Southwest shapefile. Notice all the states are now in one shapefile.** The append was successful.

In terms of file management, calling this file "Pacific_NW" could be confusing when it includes not only the Pacific Northwest states but also the southwestern states. You would likely create a new shapefile and give it a new name. Chapter 11 instructs you how to do this, but you can follow the process here.

6. **(Optional) Right-click Pacific_NW, and click Data > Export Data, and then navigate to your save folder. Rename the file, and save it. Save as type should be Shapefile.**

7. **Click File > New > Blank Map, and click OK. There's no need to save.**

> **APPENDING AND ATTRIBUTE TABLES**
> It is ideal if the attribute tables of all files in the append process are identical in terms of columns and column headings. If they are not, you can still append, but you must have at least one column in common. Search the Help menu for "Append Data Management," and read about field map control.

Clipping

Another tool you may find useful is Clip. You can clip one boundary by using the outline of another boundary. Now you can try it.

Clip tool

Example of Clipping

1. **Add States.shp and circle.shp. In the table of contents, drag circle.shp into the first position if it is not already there.** The idea is, you want to clip the states file to the boundary of the circle file.
2. **On the Geoprocessing menu, click Clip.**
3. **For Input Features, select the file to be clipped (States.shp).**
4. **For Clip Features, select the file that will serve as the clipped boundary—that is, the file that will be used to clip the first file (circle.shp).**
5. **For Output Feature Class, navigate to your save folder. Name the new file** clipped**. Click Save, and click OK.** The new file will automatically be added to ArcMap.
6. **Turn off all other shapefiles to clearly see the new clipped shapefile.**
7. **Right-click the new file, and open the attribute table. Notice it takes on properties of the States file.**

8. **Right-click the circle layer, and remove the file. Right-click the clipped layer, and remove the file. Turn the States layer back on.**

THE INTERSECT VERSUS CLIP TOOLS
The Clip tool and the Intersect tool are similar. With the Clip tool, the new shapefile takes on the attribute properties of the original layer. The Clip tool can work with only two layers at a time. Intersect is the same in using only two layers unless you have an Advanced license of ArcGIS Desktop. With an Advanced license level, the resulting attribute table can assume the attribute properties of many layers.

Dissolving

Dissolve creates larger regions out of smaller regions.

Example of Dissolve

Dissolve tool

1. **The States shapefile should already be open in ArcMap.**

2. **Right-click the file, and open the attribute table. Notice the Division column. These designations represent divisions of the US as defined by the Census Bureau. Close the attribute table.**

3. **On the Geoprocessing menu, click Dissolve.**

4. **For Input Feature Class, select States. For Output Feature Class, navigate to your save folder. Name the new file** divisions, **and click Save.**

5. **In the Dissolve Field panel, select the Division box.** There is no data in this shapefile. If there was, you could select the fields under Statistics Field(s) and sum up all the values of the states in each division.

6. **For now, leave the Dissolve field blank.**

7. **Click OK.**

8. **Close ArcMap, or to proceed to the next exercise, click File > New > Blank Map, and click OK. There's no need to save.**

CHAPTER 15

KEY CONCEPTS

creating geodatabases
working with geodatabases
working with ArcCatalog

Creating geodatabases

You may encounter many types of files while working in ArcGIS. You have been working with shapefiles, the most basic and frequently used file type in ArcGIS. Shapefiles are used extensively not only in ArcGIS but with many other GIS software programs. Another commonly used file type is the geodatabase.

In this exercise, you will explore how geodatabases work and learn about ArcCatalog. It might not get you any dates on Saturday night, but then again, there are a lot of people into geodatabases.

[*You will need to download all chapter data files from esri.com/GIS20-3.*

What is it?

A geodatabase is like a container in which you can store all files related to your GIS project. You can store multiple shapefiles, aerial photographs, spreadsheets, and many other types of files. Grouping these items together in one place is convenient because it provides one point of access for all needed files. It is also handy if you want to share these files with others. Putting everything in a geodatabase makes GIS data easier to manage and access.

If you are using only a few shapefiles for your GIS project, there is no need to convert it to a geodatabase. However, if your project begins to grow and use many files, it is worthwhile to convert the project to a geodatabase. In any case, it is good to be familiar with geodatabases, as many organizations distribute files in this format. Even though you may never find the need to create a geodatabase, it is likely that you may encounter one at some point.

Open ArcCatalog

ArcCatalog is the library system of ArcMap. You can browse, copy, delete, and organize files in ArcCatalog. ArcCatalog is also where you create file and personal geodatabases.

1. On the Windows menu, click Catalog. You can also click the Catalog button on the Standard toolbar.

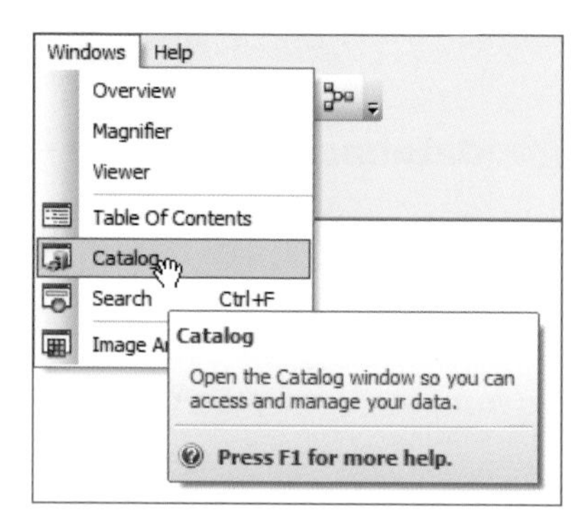

Notice a new window is open. It usually opens on the right side of the screen, but if it opens in a different location, no problem. It is still ArcCatalog. ArcCatalog looks slightly different in older versions of the software.

Explore ArcCatalog (optional)

If you are not already familiar with ArcCatalog, take this opportunity to explore it.

1. **Expand the Folder Connections folder. Notice that saved connections to folders are located here. Right-click Folder Connections** Folder Connections **, and notice you can connect to a folder using this method, instead of clicking the Connect to Folder button .**

2. **Assuming you have a file available under Folder Connections, right-click the file name and notice the menu of options.** You can create new folders as well as copy, delete, rename, and export files. These options are handy for managing files.

Create a file geodatabase

The basic process for creating a geodatabase is to first create the geodatabase, and then import or export files into the geodatabase.

1. **In ArcCatalog, under Folder Connections, navigate to your save folder. Right-click the save folder, click New, and then click File Geodatabase.**

2. **When prompted, rename the file** gdb.gdb. **When you are finished, click in the white space to deactivate the text tool.**

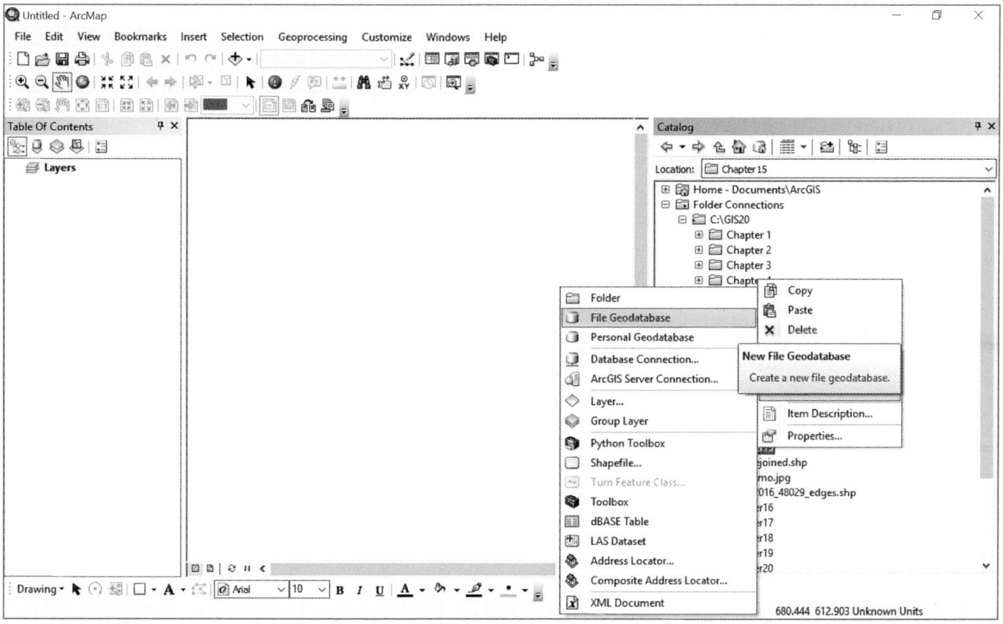

Import shapefiles to the geodatabase

1. **Right-click gdb.gdb, click Import, and click Feature Class (multiple). This will import shapefiles to the geodatabase.**

2. **In the Import Features field, navigate to your save folder, and add tl_2016_48029_edges.shp (used in chapter 8), which is the Bexar County street network. Then navigate to chapter 15, and import agejoined.shp. Click OK.** It may take a moment.

3. **Once the operation is complete, expand the geodatabase in ArcCatalog, and notice these two shapefiles have been added to the geodatabase.**

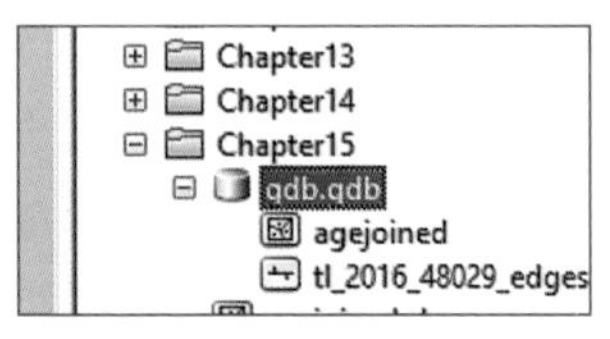

IMPORTING ONLY ONE FILE? IT'S A LITTLE DIFFERENT FROM IMPORTING MULTIPLE FILES

If you are importing only one file into the geodatabase, a field named Output Feature Class requires an input. What it wants is simply what you want to name this file within the geodatabase. For example, you might name the tl_2016_48929_edges shapefile Streets, instead of the long given name from the Census. This field does not require the full file path, only the name.

Export aerial photographs to the geodatabase

The import function does not work with aerial photographs, nor does it work with Excel spreadsheets. Instead, you must first navigate to those files, and then export them to the geodatabase. You'll use Alamo.jpg from chapter 17, but you can use it simply to have a photograph to import. Alamo.jpg is located on the book resource web page, in chapter 15 or chapter 17 of your save folder.

1. **In ArcCatalog, navigate to your save folder, chapter 15 (assuming you downloaded all the files from the book resource web page), to the georeferenced Alamo aerial photograph (Alamo.jpg).**

2. **Right-click the file, click Export, and then click Raster to Different Format.** The Input Features field should already contain the file to export to the geodatabase.

3. **In the Output Location field, navigate to gdb.gdb. Double-click the file, and type a new name. Call it** Alamo. **(A file name with spaces will generate an error, and the operation will fail.) Ensure that "Save as type" is Raster Datasets. Click Save.**

4. **Leave all the other options, and click OK.** It will take a minute to process.

5. **Once it finishes, in ArcCatalog double-click gdb.gdb, and notice the file has been added to the geodatabase.** It will also open in ArcMap.

   TIP *You can also drag files from ArcCatalog into the table of contents to easily open them.*

Export an Excel table to the geodatabase

1. **In ArcCatalog, navigate to chapter 15 in your save folder, and double-click Agencies.xls.**

2. **Right-click the worksheet Organizations$, click Export, and then click To Geodatabase (single).** The Input Features field should already contain the file to export to the geodatabase.

3. **In the Output Location field, navigate to gdb.gdb. Do not give it a new name. Click Add.**

4. **In the Output Table field, type a new name. Call it** Addresses. **You can name it anything you like, but remember, if the file has spaces, the import will fail. Click OK.** The import will take a minute to process.

5. **Once it finishes, in ArcCatalog double-click gdb.gdb, and notice the file has been added to the geodatabase.** It will also open in ArcMap, but you will not be able to see it because it does not have any geographic attributes.

6. **Close ArcMap, or to proceed to the next exercise, click File > New > Blank Map, and click OK. There's no need to save.**

GEODATABASES SERVED THREE WAYS

ArcGIS supports three types of geodatabases: file (.gdb), personal (.mdb), and ArcSDE (.sde).

For one person, or small workgroup projects, the file geodatabase is the best choice. It has no size limitations. The personal geodatabase is still used, but this older file type is limited to 2 GB storage space.

The ArcSDE geodatabase (SDE stands for Spatial Database Engine) is for large workgroups. It is not accessible with an ArcGIS Desktop Basic license. This type of geodatabase is created and administered by a GIS manager. An SDE has some big advantages. Files can be locked so they cannot be edited by different people, helping to maintain data integrity. Also, an SDE can store and draw raster data, such as aerial photographs, quite efficiently. Another key advantage is that multiple people can work with, and even edit, data simultaneously, unlike in file and personal geodatabases.

CHAPTER 16

KEY CONCEPT
using the Spatial Join tool

Joining boundaries

In chapter 5, you joined a data table to a shapefile. In this chapter, you will perform a spatial join, which involves combining two shapefiles (and their data tables) into one. For example, what if you wanted to figure out which census tracts are in which neighborhoods in your city? One way is to click each tract on a map and write down the neighborhood it's in. But that would take far too long. Instead, you can perform a quick spatial join to combine tract and neighborhood shapefiles. The result will be an attribute table that includes both tract and neighborhood data.

Spatial joining is an extremely powerful tool, yet it is the kind of thing you don't know you need until the minute you realize you do. It will save you a lot of work.

In this exercise, you'll create a data table, which will tell you which counties each city falls within. Cities, after all, can be in multiple counties. Using this technique, you'll be able to determine, for example, that Boaz City spans DeKalb, Marshall, and Etowah Counties.

You will need to download all chapter data files from esri.com/GIS20-3.

Add shapefiles

1. **Open ArcMap. Click Add Data to add AlabamaCounties.shp and tl_2016_01)_place.shp, accessible from the book resource web page, in chapter 16 of your save folder.** Just so you know, you can add both shapefiles simultaneously by pressing and holding the Ctrl key and clicking on each file before clicking Add.

TWO SHAPEFILES BETTER THAN ONE?
Even though both shapefiles are added, the place shapefile might be the bottom layer, because it's blocked out by the counties, making it look as if you have only one shapefile open. You can pull the place shapefile into first position by clicking the List By Drawing Order button in the table of contents, and then dragging the layer into first position. At that point, you should see two open shapefiles.

Activate the Spatial Join tool

1. **Click the ArcToolbox button to open ArcToolbox.**
2. **Expand the Analysis Tools toolbox and then the Overlay toolset.**
3. **Double-click the Spatial Join tool.**

Fill in spatial join options

1. **For Target Features, select the shapefile to which you want to append the data.** You will use the place shapefile, tl_2016_01_place.shp.
2. **For Join Features, select the file you want to append.** You will use the county shapefile, AlabamaCounties.shp.
3. **For Output Feature Class, navigate to your save folder. Name the file** spatialjoin**. Click Save.**
4. **In the Join Operation field, leave the default option, Join_One_to_One.** Here's where things get a little tricky. In the Field Map section, you must indicate you want each county's name to appear in the results column, separated by a comma.
5. **Scroll toward the bottom, right-click Name_1 (Text), and click Properties.** Be careful, there is a Name field, but the field you want is called Name_1. This distinction is important because a city can span multiple counties. You want the results to indicate each county that the city falls within, even if it falls within multiple counties.

6. **Change the length from 100 to 254, the maximum characters allowed, by double-clicking on 100. For Merge Rule, select Join (do not leave as first), and type a comma for the Delimiter field. Click OK.**

   NOTE: *If you had not selected Join for the Merge Rule type, and instead left it as first, only the first encountered county would be listed, regardless of whether a city spanned multiple counties.*

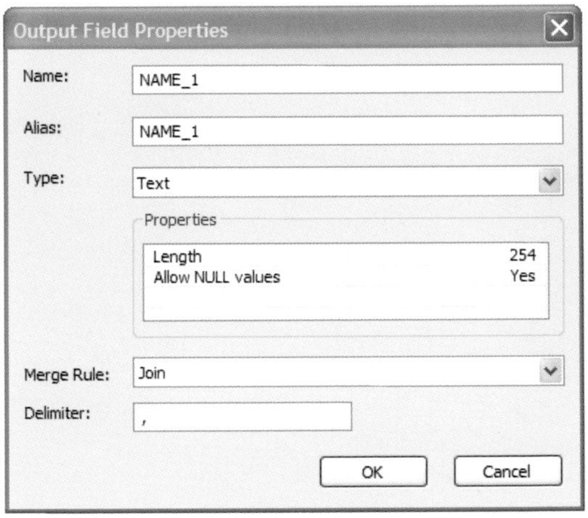

   Output Field Properties
   Name: NAME_1
   Alias: NAME_1
   Type: Text
   Properties
   Length 254
   Allow NULL values Yes
   Merge Rule: Join
   Delimiter: ,
   OK Cancel

   The next step is important.

7. **Scroll down on the pop-up window and notice the For Match Option drop-down list. Select Intersect because you want to know which cities intersect which counties.**

8. **Click OK twice, and wait for the green check mark to pop up in the lower-right corner.** The operation may take a minute. The new spatial join shapefile will be added to the table of contents.

Open the attribute table, and view the results

1. **In the table of contents, right-click the new shapefile name, and open the attribute table.** Notice the Name_1 column now indicates which counties the cities fall in. The city is listed in the Name column toward the beginning of the table. But you must scroll to the right to see the Name_1 column and the county information.

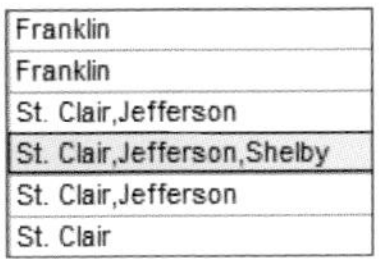

Franklin
Franklin
St. Clair,Jefferson
St. Clair,Jefferson,Shelby
St. Clair,Jefferson
St. Clair

2. **Close ArcMap, or to proceed to the next exercise, click File > New > Blank Map, and click OK. There's no need to save.**

CHAPTER 17

KEY CONCEPTS
importing
georeferencing

Working with aerial photography

Aerial photographs are great for displaying a snapshot of what is happening on the ground. They are not considered "smart maps" because you cannot click the map and get information about a particular feature.

Aerial photography, a type of raster data, can be especially useful when combined with shapefiles or other vector data. For example, utility companies may want to view a picture of a five-block area, and then overlay (in shapefile format) the aerial photograph with utility poles (points perhaps collected from a GPS) and associated information about those poles such as electricity generated or malfunctioning power lines.

This type of information changes constantly and can easily be updated in a shapefile, but not in an aerial photograph. Therefore, learning to work with both types of files, especially in combination, is useful.

You will need to download all chapter data files from esri.com/GIS20-3.

Not going to lie, this exercise is the hardest chapter in the book. In this exercise, you will focus on aerial photographs; however, these techniques would also work well with a scanned map.

You will import an aerial photograph of the area around the Alamo, a US national historic landmark and the site of a key battle, now located in downtown San Antonio, Texas. You will then learn how to overlay a street network on top of the image. You will learn how raster and vector data work together by georeferencing raster data.

Add the shapefile

1. **Add tl_2016_48029_edges.shp by clicking Add Data (assuming you downloaded the files, from chapter 17 in your save folder.** This shapefile contains streets in Bexar County, Texas, where the Alamo is located.

   TIP ***For this type of project, always add the shapefile before adding the aerial photograph.***

Add an image to your map

1. **Click Add Data again, and navigate to the aerial photograph Alamo.jpg.**
2. **When a dialog box appears warning of an "Unknown Spatial Reference," click OK.** The image will be added to the table of contents but will not be visible in data view yet. Three bands will be added under the photograph.

NO SPATIAL REFERENCE: WHY THIS IS A PROBLEM
In this exercise, you want to display the street file on top of the image file. Right now, the image file has no spatial reference so the program does not know where to put it in relation to the street file.

Move between the two layers

1. **To view each file, in the table of contents, right-click the aerial photograph's layer name, and click Zoom To Layer.**
2. **To view the streets, do the same thing but with the street layer.**

Identify four intersections in the aerial photograph

You must identify at least 64 points—kidding! You need identify only three points (the more the better, though) on the image. These points will serve as anchor points for spatial referencing. To make your task easier, four intersections have been identified and labeled in the figure:

- Alamo Plaza & E Houston Street (northwest corner)
- Alamo Plaza & E Crockett Street (southwest corner)
- E Crockett & Bonham (southeast corner)
- E Houston & Bonham (northeast corner)

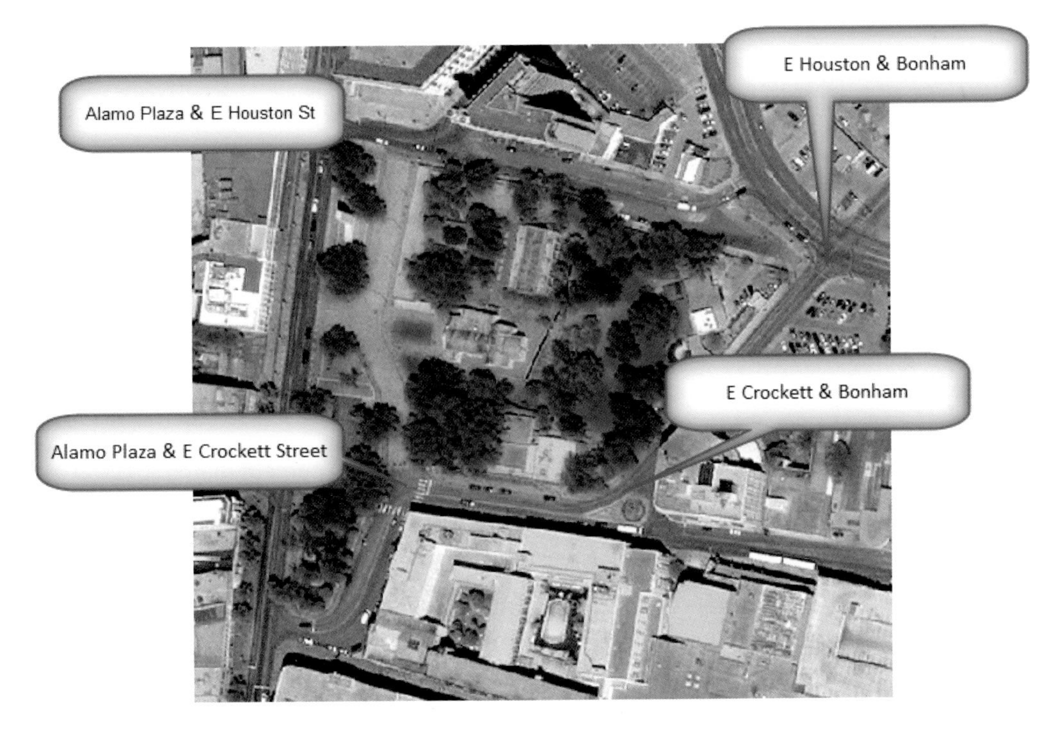

Identify the first intersection on the street network in ArcMap

1. **In the table of contents, right-click the street network layer, and click Zoom To Layer. Ensure that you are looking at the street network before proceeding.**

2. **Click the Find tool, which looks like a pair of binoculars, next to the Identify tool.**

3. **Click the Locations tab.**

NOTE: *The Choose a Locator field should say Esri World Geocoder. If it does not, you must update to the latest version of ArcGIS.*

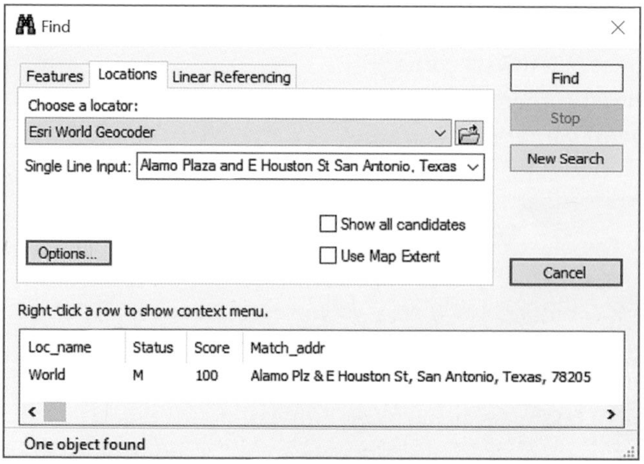

4. **In the Single Line Input field, type** Alamo Plaza and E Houston St San Antonio, Texas**. Click Find.** The results will display in the box at the bottom.

5. **Click either of the presented intersections.** Both are a 100 percent match so you might as well select the first one. Notice the spot on the map flashes, but the view is still too far out to see much.

6. **Right-click the first intersection, and click Add Point ⊙.** A point is placed at that intersection so it is easy to see.

7. **Right-click again on the same intersection, and click Zoom To, which will zoom to the point. Click Cancel to close the dialog box.**

TIP ***If Add Point is unavailable, make sure you are in data view. If you are not in data view, go to View > Data View on the menu bar to switch from layout view to data view.***

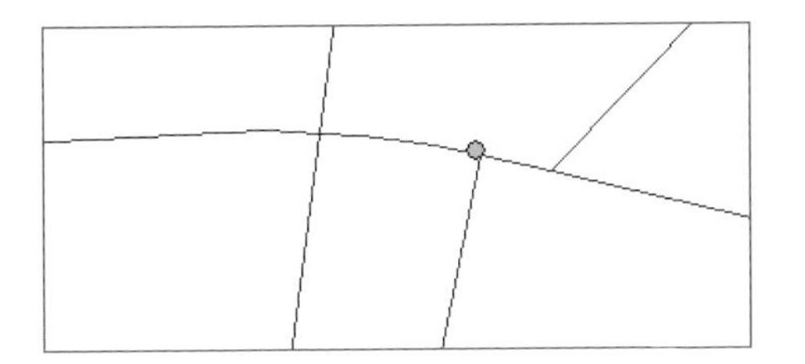

Fit to display

1. **On the Customize menu, click Toolbars, and then click Georeferencing.**

2. **Click the Georeferencing arrow** Georeferencing ▾, **and select Fit to Display.** Then something magical happens. Both the aerial photograph and streets are displayed in the same view, but alas, they are not aligned. This work still lies ahead. It's important to be able to see the street layer easily.

3. **In the table of contents, right-click the street layer name, click Properties, and then click the Symbology tab. Click the symbol swatch, and change the streets to a color and width that are easier to see.** Highway is a good option for symbolization.

Add control points

Adding control points allows you to align the image with the shapefile layer.

1. **Turn the street layer off, and view the image presented earlier in this chapter (under "Identify four intersections in this aerial photograph") to figure out where the intersection of Alamo Plaza and E Houston Street is on the aerial photograph.** Hint: It is in between two white cars in the northwest corner of the map.

2. **Turn the street layer back on.**

3. **On the Georeferencing toolbar, click the Add Control Points tool .**

4. **First, with the Add Control Points tool active, click the image where the intersection is located. Drag the line that appears to the intersection on the street network. It must be done in this order—click the image first and then the shapefile.** The intersections should line up somewhat. This is control point 1.

5. **If you haven't deleted the dot you placed earlier, click the default pointer (Select Elements tool) , click the dot, and click Delete on your keyboard.** Now, we're having fun!

WHAT HAPPENS IF I MESS UP?
The View Link table catalogs all control points. If you accidentally click the wrong point, no problem. Click the Delete Link button , select the point, and delete it. Alternatively, on the Georeferencing toolbar, you can click the View Link Table button . Click the point you want to delete in the table, and click Delete on your keyboard. The point is deleted. You can then close the table or keep it open.

6. **Add another control point for the intersection in the southwest corner, Alamo Plaza and E Crockett Street. Remember, click the image first and then the shapefile.** The intersections should line up. This is control point 2.

NOTE: *Clear the Use Map Extent check box because otherwise it will limit the search to the streets visible on your screen.*

7. **Do the same thing for the other two intersections.** Throughout this process, you will likely need to zoom in and out to see more clearly. The mouse scroll wheel is great for zooming.

View the View Link table

1. **After all four points are placed (or more if you're getting into it), on the Georeferencing toolbar, click the View Link Table button . Scroll right, and notice the residual column at the end.** This column indicates how closely the two files are aligned. A total score is given in the upper-right corner in the Total RMS Error field. RMS stands for root mean squared. The closer this number is to zero the better.

   Total RMS Error: Forward: 19.51

   You won't have good guidelines on what the RMS should be. In a perfect world, the RMS would be zero. Does perfection happen in real life? Never. Do the best you can.

2. **Close the table.**

Update georeferencing

1. **Once you have finished adding points and working with the data, on the Georeferencing toolbar, click the Georeferencing button** Georeferencing ▾ **.**

2. **Click Update Georeferencing.** Now the image can be used in conjunction with other shapefiles for this area.

3. **Crack open a tasty beverage. You deserve it.**

4. **Close ArcMap, or to proceed to the next exercise, click File > New > Blank Map, and click OK. There's no need to save.**

I HAVE A SCANNED MAP—HOW CAN I TURN IT INTO A SHAPEFILE?
It is impossible to turn a map into a shapefile. Doing so would be magic. You can follow the steps in this exercise, except substitute a scanned map for the aerial photograph. Using a scanned map would let you digitize it and overlay other layers on top of it. You could also use editing tools to create a new shapefile and sketch a boundary of something like a target area. For more on this subject, see chapter 11.

CHAPTER 18

KEY CONCEPTS

creating reports in ArcGIS

exporting reports in ArcGIS

Creating reports

If you use GIS to provide information to other people, you may find creating reports a useful feature. Reports can give your readers a lot of information and provide credibility for the map's data. Report data can either be included as a part of your map's layout (if it is a small amount of data) or attached as a technical addendum.

In this exercise, you will create and export a basic data report.

You will need to download all chapter data files from esri.com/GIS20-3.

Add shapefiles

1. **Click Add Data to add agejoined.shp.** You will create a report for a small selection of counties, not all counties.

2. **Zoom in fairly close to your map, displaying about 10 counties, which will be the basis of your report.**

Open the attribute table

1. **To refresh your memory about the types of data you have in this shapefile, in the table of contents, right-click agejoined, and click Open Attribute Table.**

2. **Scroll far to the right, and notice that each county has an entry for total population and percentage of seniors.** These columns will form the basis of your report. If you don't have these columns, you can grab them from the data files on the book resource web page, in \GIS20\Chapter18.

Create a report

1. **From within the attribute table, click the Table Options button (upper-left corner), click Reports, and then click Create Report.**

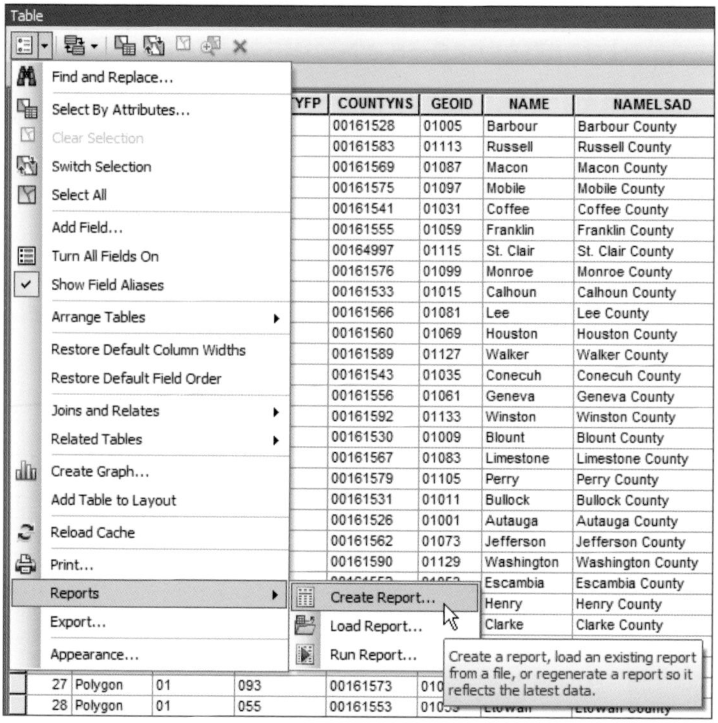

2. **Under Available Fields, scroll down and select County. Click the right arrow to deposit it into the Report Fields section. Do the same thing with the Seniors column.** The report will have two columns.

3. **Click the Dataset Options button, select Visible Extent, and then click OK.** This setup will create a report for only those counties displayed in the map frame.

4. **Click Next twice.**

5. **When you get to the field-sorting options, click the arrow under Fields, and select the Seniors column. Click the arrow under Sort, and select Descending.** This option will put higher percentages at the top of the column.

6. **Click the Summary Options button.** Summary Options calculates the average, which is often helpful. It also does a few other calculations.

7. **Select the Avg box, click OK, and then click Next.**

8. **For the report layout, leave the defaults, and click Next.**

9. **For this step, review the styles.** Good options for styles are Chicago, New York, and Simple.

10. **Select Chicago, and click Next.**

11. **Type** Percentage of Seniors **as the title of the report, and then click Finish.** Your data will display in report format. But the column headers may be misaligned with the numbers.

12. **To align the column headers, click the Edit the Report Properties button** Edit... **, click once on the second Seniors label, and then click the Align Left button ≡ on top.**

13. To make it all work, click the Run Report (F5) button in the upper-left corner.

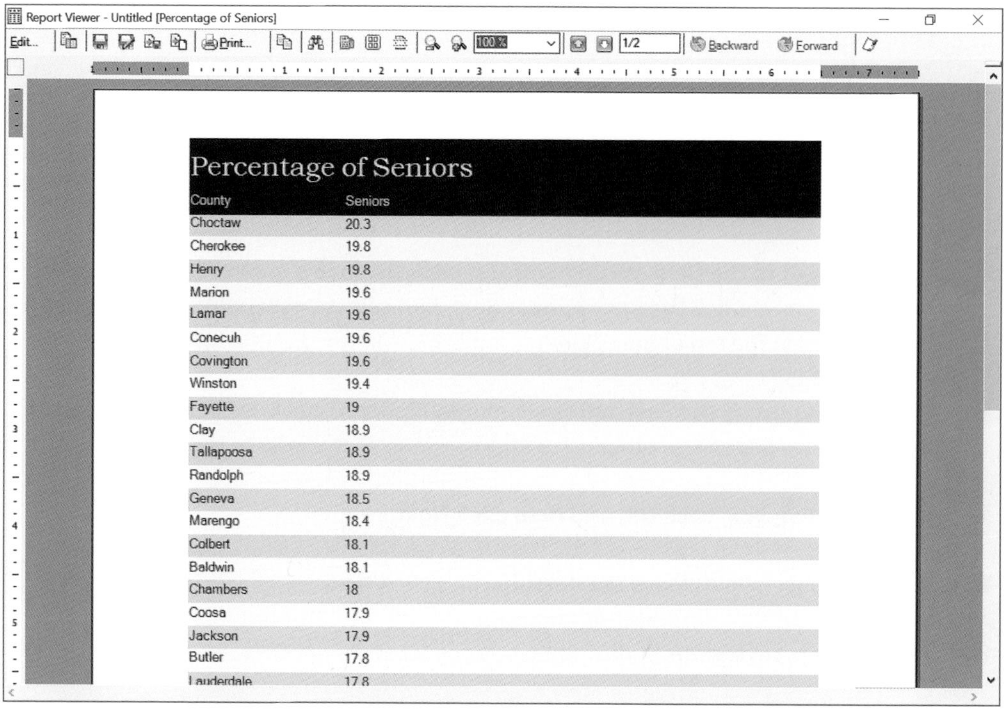

Report Viewer - Untitled [Percentage of Seniors]

Percentage of Seniors

County	Seniors
Choctaw	20.3
Cherokee	19.8
Henry	19.8
Marion	19.6
Lamar	19.6
Conecuh	19.6
Covington	19.6
Winston	19.4
Fayette	19
Clay	18.9
Tallapoosa	18.9
Randolph	18.9
Geneva	18.5
Marengo	18.4
Colbert	18.1
Baldwin	18.1
Chambers	18
Coosa	17.9
Jackson	17.9
Butler	17.8
Lauderdale	17.8

Export report

You may want to export your report and attach it to your map, or you may want to export it to Excel for further analysis.

1. **On the toolbar at the top of the report, click the Export Report to File button .**

2. **Click the Export Format arrow, and notice several options. Select Portable Document Format (PDF).**

3. **In the File Name field, click the ellipsis, and navigate to where you want to save the file. Name it** report**. Click Save, and click OK.** You have saved your report in PDF format.

4. **Close ArcMap, or to proceed to the next exercise, click File > New > Blank Map, and click OK. There's no need to save.**

ATTRIBUTE QUERIES IN REPORTS

The ability to write an attribute query and have the results of that query make up the report is a useful report feature available in ArcGIS. When you select the columns to include, if you click Dataset Options, you can select certain records you want to include in the report. You also have the option to select Definition Query as the option, and a query wizard will open. You can write an attribute query that will select records on the basis of your own criteria. See chapter 12 for more on writing attribute queries.

CHAPTER 19

KEY CONCEPT
using layer, map, and geoprocessing packages

Sharing work

It is easier than ever to share maps and all the pieces that make up those maps. You can share an entire workspace or just a few layers. With the package functionality available in ArcGIS, it is easy to send layers and projects via email. It is also easy to upload data to ArcGIS Online, which is covered in chapter 20.

Ever thought it might be a good idea to figure out how to move these files between your desktop, laptop, and your coworkers? Well, this chapter's for you. In this exercise, you will learn how to package maps, layers, and tools for the purpose of sharing your work with others.

This chapter focuses on offline saving and sharing options. Chapter 20 shows you how to share your work to ArcGIS Online.

You will need to download all chapter data files from esri.com/GIS20-3.

Open the project

1. **Open ArcMap. On the File menu, click Open, and open seniors.mxd, which was created in chapter 6 and is also accessible in chapter 19 from the book resource web page.**

Map packages

Save the map package to your desktop

This method saves the map package (.mpk) to your desktop. You can then email the map package or save it to a flash drive. It contains all the files for this project. The end user will see exactly what you see in ArcMap.

1. **On the File menu, click File > Share As > Map Package.**

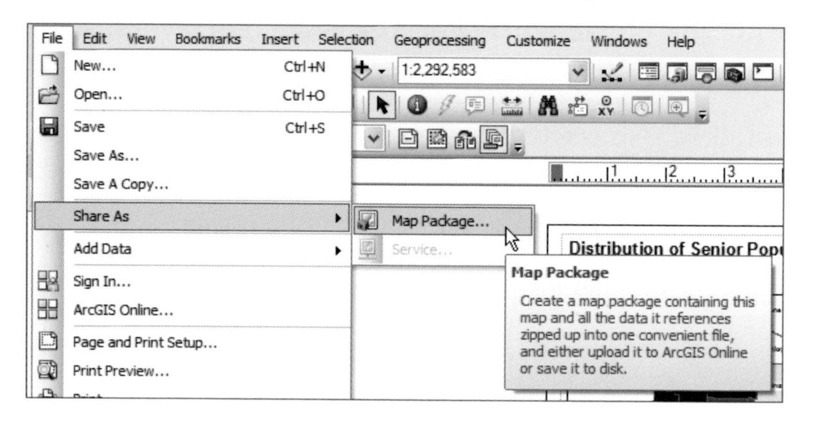

2. **Select Save Package to File, navigate to your save folder, leave the name as Seniors.mpk, and click Save.**

3. **On the left, select Item Description, and fill in the required fields. For the summary, type** Senior Population. **For tags, type** seniors. **Add a description, even though it doesn't say one is required. Type** seniors. **Select the box to update missing metadata based on item description.**

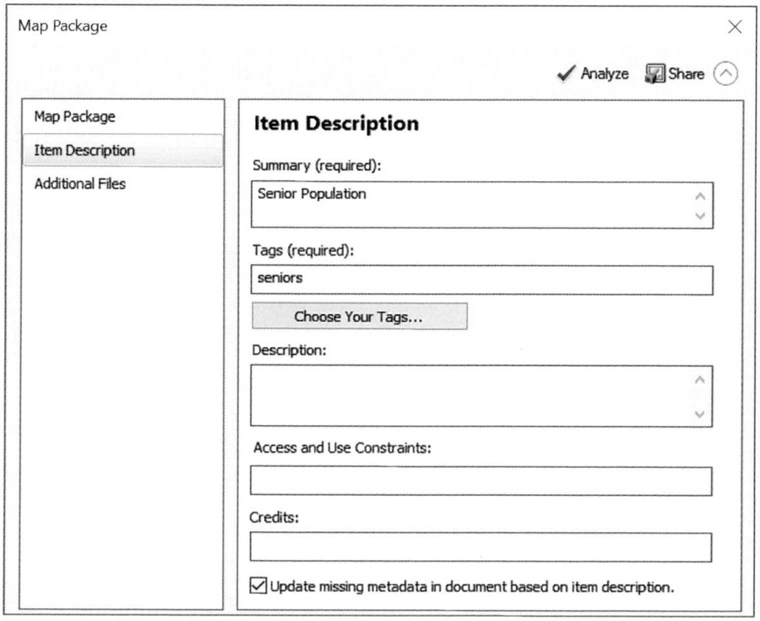

4. **In the upper-right corner, click the Share button. Click OK when the operation succeeds.** A new map package file is created on your desktop. This package contains all the files that make up this project.

5. **Close ArcMap. Navigate to the seniors.mpk file in your save folder. Double-click the file.** It will open in ArcMap.

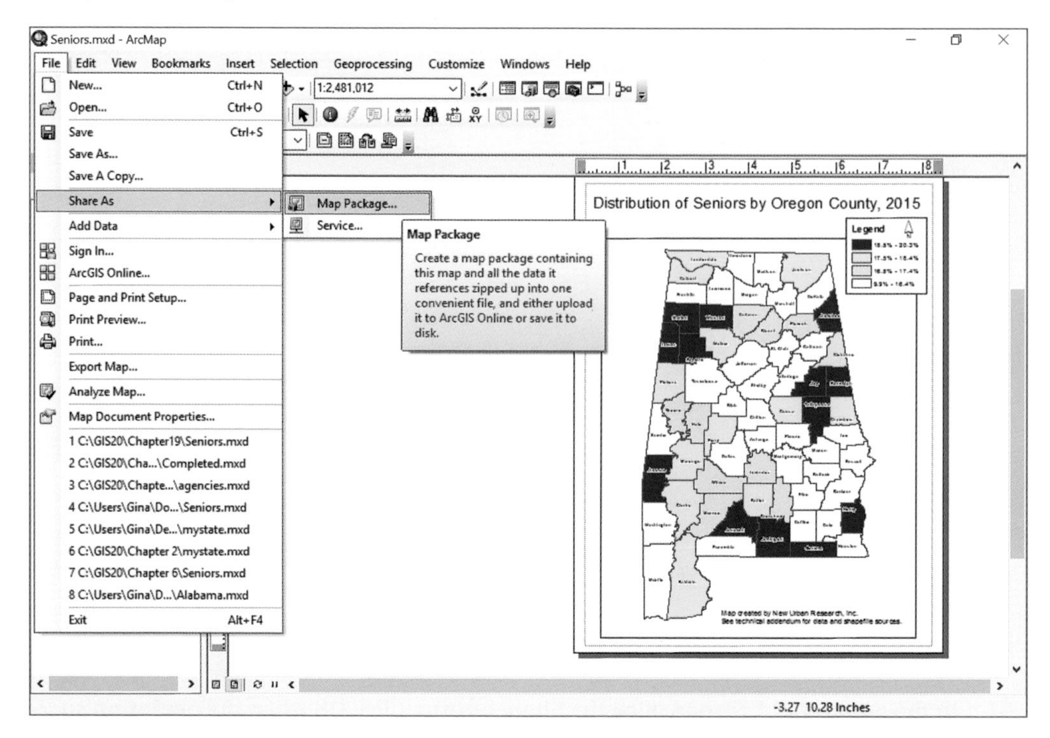

Layer packages

Layer packages (.lpk) allow you to save individual layers, not the entire workspace. This type of package will save the layer (including color shading and labeling) along with the underlying shapefile, making the layer package a handy option.

Save a layer package to your desktop

1. **In the table of contents, right-click the agejoined layer name, and click Properties.**

2. **Click the General tab. In the Description panel, type** seniors **(there must be a description). Click OK.**

3. **In the table of contents, right-click the agejoined layer name, and click Create Layer Package.**

4. **On the left, select Item Description, and fill in the required fields. For the summary, type** seniors. **For tags, type** seniors. **And for Description, type** seniors.

5. **On the left, click Layer Package. Select "Save package to file," navigate to your save folder, leave the name as agejoined.lpk, and click Save.**

6. **In the upper-right corner, click the Share button. Click OK once the operation succeeds.** A new layer package file is created on your desktop. This package contains this layer, plus its data.

7. **Close ArcMap. You do not need to save anything. Navigate to the agejoined.lpk file in your save folder. Double-click the file.** It will open in ArcMap. It may take a moment.

Geoprocessing packages

Geoprocessing packages (.gpk) allow you to share your workflow with others. Chapter 14 covered some basics of geoprocessing. Imagine being able to zip that whole edit session and send it to someone else. That's what a geoprocessing package does. All resources (models, scripts, data, layers, and files) needed to run the tools are included in the package.

Save geoprocessing package to your desktop

Now you can use the clip from using the Clip tool to create a geoprocessing package.

1. **Open states.shp and circle.shp again from chapter 14, also in chapter 19, of your save folder. In the table of contents, drag circle.shp into the first position.** You want to clip the states file to the boundary of the circle file.

2. **On the Geoprocessing menu, select Clip.**

3. **For Input Features, select the file to be clipped (states.shp).**

4. **For Clip Features, select the file that will serve as the clipped boundary, meaning the file that will be used to clip the first file (circle.shp).**

5. **For Output Feature Class, navigate to your save folder, name the new file** clippedagain, **click Save, and then click OK.** The new file will automatically be added to ArcMap.

6. **On the Geoprocessing menu, click Current Session, and expand to Results.**

7. **In the Results window, right-click Clip, click Share As, and then click Geoprocessing Package.**

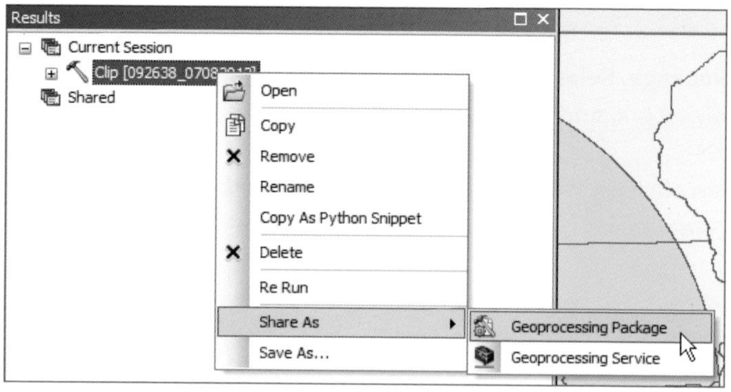

TIP *You can also publish as a service if your organization maintains an ArcMap server.*

8. **Select Save Package to File. Click the Browse button, and navigate to your save folder, leave the name as Clip.gpk, and click Save.**

9. **Clear the Include Enterprise Geodatabase check box. Click Share. Click OK when the results are successful.**

Rerun the geoprocessing package

1. **On the File menu, click New > Blank Map, and click OK. This opens a clean workspace.**

2. **Open ArcCatalog, navigate to your save folder, and notice a new file named Clip.gpk.**

3. **Drag Clip.gpk into the empty mapping window. In the table of contents, click the plus sign next to the Clip layer to expand the layer.** The three pertinent shapefiles are now available in the table of contents.

4. **Close ArcCatalog.**

5. **To rerun these results, remove the clipped file so you can re-create it by rerunning the geoprocessing. Right-click the clippedagain shapefile, and remove it.**

6. **On the Geoprocessing menu, click Results to open the Results window. Click the plus sign next to Clip to expand it. Right-click the second Clip link, and click Re Run.** Give the tool a moment to work. After it is finished, the newly clipped shapefile will be added to the table of contents. Hooray!

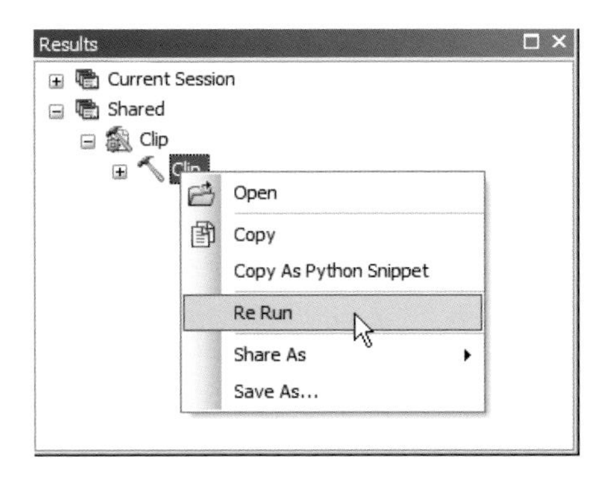

7. **Expand all the menus in the Results window, and notice all the information available about the operation. Close the Results window.**

8. **Close ArcMap, or to proceed to the next exercise, click File > New > Blank Map, and click OK. There's no need to save.**

LOCATOR PACKAGES

Another package tool provides the ability to create address locator packages (.apk) and share them with others so the packages can be used during their geocoding edit sessions. A lot can go into creating an address locator, especially if it is a composite address locator with a comprehensive alias table. This tool is in ArcToolbox > Data Management Tools > Package > Package Locator.

CHAPTER 20

KEY CONCEPTS

exploring ArcGIS Online
creating geoenabled PDFs

Publishing maps

You have several good options for publishing and sharing maps. ArcGIS Online provides a space for users to upload and download maps, data, packages, and all sorts of other useful stuff. Think of it as a community for GIS users. Geoenabled PDFs are another useful way to publish your maps (in print form). The "geoenabled" part allows you to package the underlying data along with the maps, which in turn allows users to click the map and access attribute data.

In this exercise, you will explore ArcGIS Online and create a geoenabled PDF file. So what's so cool about this? Everything. It's how you get your maps into the world and how others can take them in and behold.

[*You will need to download all chapter data files from esri.com/GIS20-3.*

Create an ArcGIS Online Public Account

The ArcGIS℠ Online Public Account, which you can create for personal use, is a free account and is required to access many features in ArcGIS Online.

1. **Open an internet browser, and navigate to http://www.arcgis.com.**

2. **Click the Sign In link, and then click the Create A Public Account button.**

3. **Fill in the required fields. You must review and accept the terms of use to create an account.**

4. **(Optional) Feel free to write a short description about yourself, make your profile public or private, and set language and location preferences. You can also skip this step.**

> **GROUPS: PUBLIC AND PRIVATE**
> ArcGIS Online provides the valuable resource of allowing users to create groups that can be public or private. For private groups, members can join by invitation only, and no search will identify the group. This privacy is great for internal workgroups. Public groups, on the other hand, will turn up in search results, and anyone can apply to join the group. You can search for groups to join by clicking the Group button and then searching for a group.

ArcGIS Online

Explore the ArcGIS Online Gallery

1. **Click the Gallery link.**

2. **In the search field in the upper-right corner, type** Median Income**, and click the Search button. Select any map that interests you, preferably of the whole US and not a Premium Content layer.** Content maps require credits (if you still have credits, you can use the premium maps, of course). This exercise assumes you don't have credits, and any map will do. Toward the bottom of the page, there's a map called Median Household Income 2011—this map is the one used in this exercise.

3. **Click the Open in Map Viewer button.**

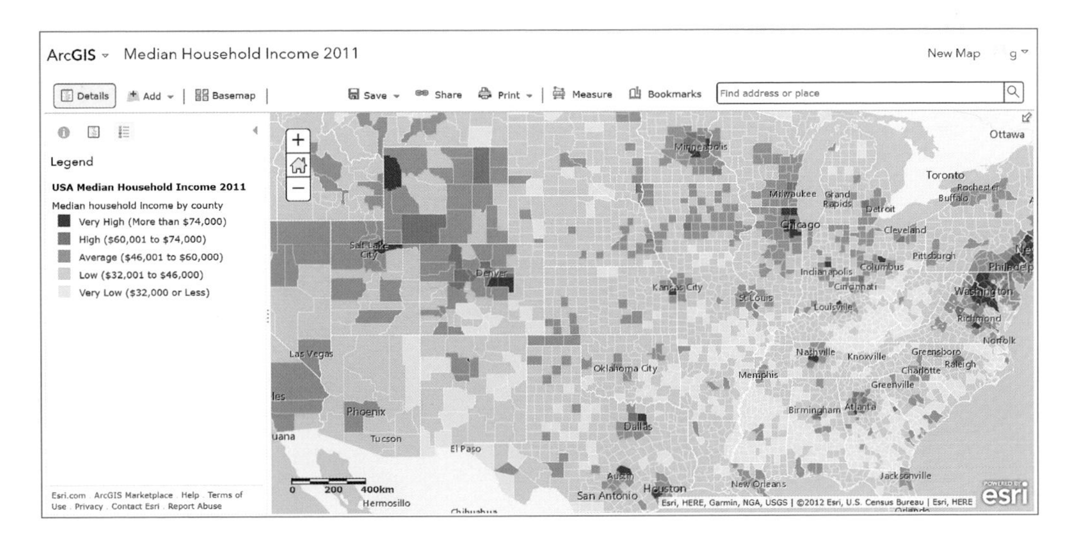

Find a location and geographic scale

1. **In the upper-right corner, in the Search field, type your city's name. For this exercise, Portland, Oregon, is used.** The map opens, and on the left, the Legend pane shows a range of median household income in 2011.

2. **Practice zooming in and out by using your mouse scroll wheel (or the Zoom In and Zoom Out tools).** The geography changes from block group to tract to county and to state.

3. **Have some fun, and type your own address, and then see what the median income of your block group is compared to everyone else's.**

Access data

These kinds of maps are just static, but you can click on individual polygons and see data (attribute tables are not available in ArcGIS Online maps). You will not use the Add Data button for this step.

1. **Click on any area of the map. Notice an information box appears, and the geography is highlighted in bright blue.**

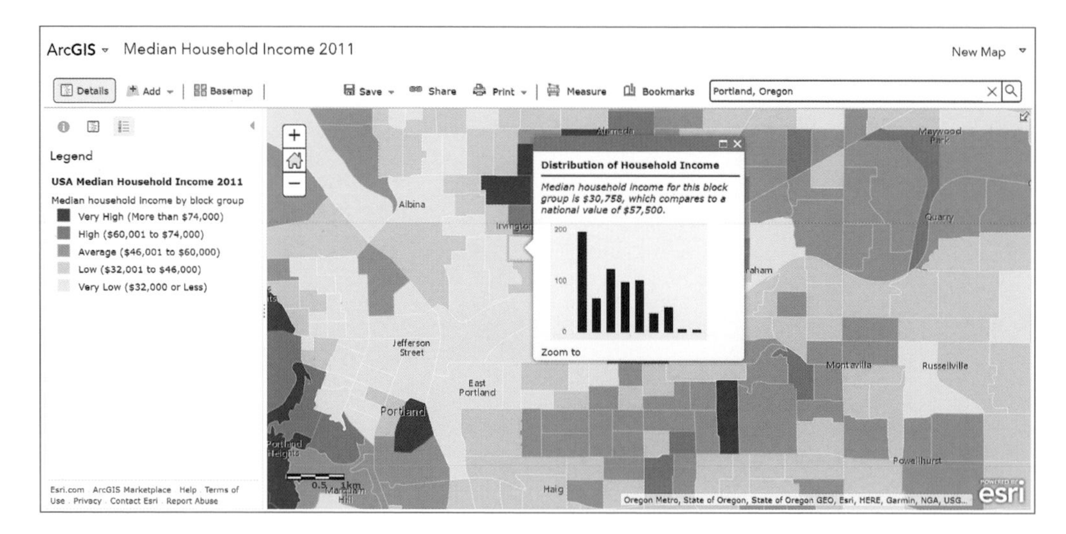

Save your map

1. **At the top above the map, click Save > Save As.**

2. **Fill in the options, and click Save Map.**

Share using social media

1. **Click the Share button.**

2. **Select the check box to allow everyone to view your map.**

3. **If you have a Facebook or Twitter account, try posting your map directly from ArcGIS Online.**

EMBED YOUR MAP IN A WEBSITE

On the Share menu, you have the option to embed your map in a website. This link provides the code to create an iframe. To make it work, you must copy the code into a blank HTML page that you can then upload to a server and publish to your website. That's all you have to do to get a map onto your website.

Share by web app

1. **In the Share window, click the Create a Web App button. This feature allows you to make and store maps on Esri servers. You can store up to 2 GB for free. After that, you must purchase a plan.**

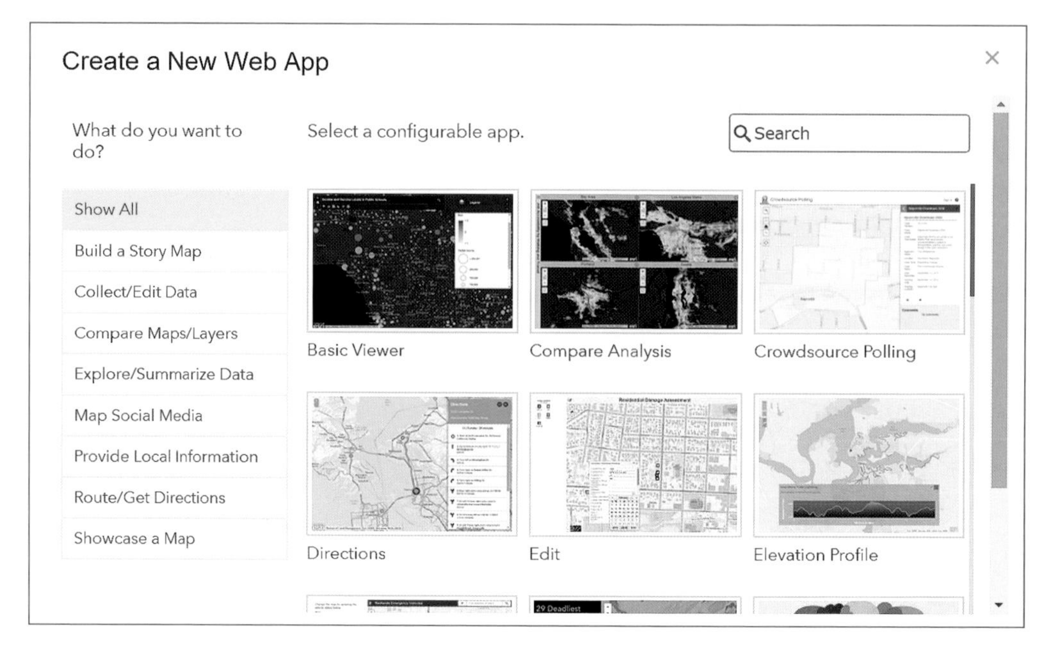

2. **Click Basic Viewer (the first square).**

3. **Use the scroll bar on the right side of the pop-up window to scroll down. Click the Preview button to open a new browser tab with the map.**

4. **Notice all the great options, including the ability to move the map, zoom, and type an address, as well as many other features. Once you are finished looking around, close the tab.** Don't worry—you won't lose anything. You should still have a browser tab open for ArcGIS Online.

5. **Click Create Web App, and then click Done.** On the web app configuration page, you can change the app properties, such as theme, color, and map title, by clicking the General, Theme, Options, and Search buttons in the upper-left corner.

6. **Feel free to play around with these settings.**

7. **Click the Launch button, which is at the bottom of the map—you will need to use the scroll bar.** You have now created your very own map on the internet, which can be shared.

8. **Click the Share button, and share by copying and pasting the map link.**

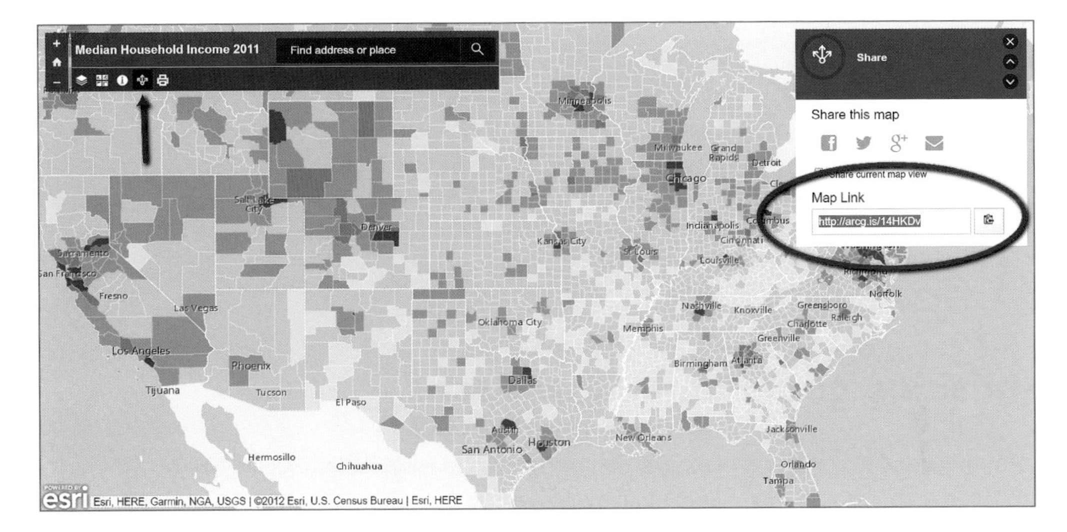

Upload to ArcGIS Online

1. **Open a new internet tab, and go back to** arcgis.com. **You should still be logged in. Click the Content tab, and then click Add Item > From My Computer.**

2. **Click Choose File, and navigate to seniors.mpk in chapter 20 of your save folder. For tags, type** senior. Tags are what will enable you (and others) to find your file in ArcGIS Online.

3. **Click Add Item (you might need to scroll to the right to see the button).**

TIP ***Different file types provide slightly different options. If the option to open in ArcMap is available, it will open automatically in ArcMap. You can use this option, and skip the saving to desktop option.***

4. **On the upper right, click the Share button, select the check box to share it with everyone, and click OK.** It's now shared with the world, and others could now click the Download button, download your map package, double-click on it, and they would see exactly what you see in ArcMap.

BASEMAPS IN ARCMAP

You might have noticed the ability to add basemaps to ArcGIS when clicking the Add Data button if you clicked the little down arrow on the button. The kind of maps you've been working with in this exercise are basemaps.

Add Data...
Add Basemap...
Add Data From ArcGIS Online...

Basemaps do not contain attribute information and are not shapefiles. They are visual images of maps, but you cannot join to them or retrieve attribute data from them. They are helpful to give contextual information to your map but are not too helpful for analysis.

ARCGIS ONLINE: FREE OR FEE?

ArcGIS Online is free to use. You can store up to 2 GB for free with just a personal ArcGIS Online Public Account. After that, an organization subscription is necessary. For more information, go to http://www.esri.com > Products > ArcGIS Online.

Publishing geoenabled PDF maps

Another good way to share maps is to export them as a geoenabled PDF. In Adobe® Acrobat® Reader 6.0 and higher, you can turn individual map layers on and off. With Adobe Reader 9.0, you can use many new tools that work with ArcMap. The following steps work with either version, but it is a good idea to upgrade if you haven't already. Adobe Acrobat Reader is free software.

Open the project

1. **Open ArcMap. On the File menu, click Open. Navigate to seniors.mxd in your save folder.**

Modify the attribute table

Select only a few columns to bring over with the PDF since you don't need all the columns.

1. **In the table of contents, right-click the layer name, and click Properties.**
2. **Click the Fields tab, and notice all columns have a check mark next to them.** If you left all the fields like this, all these columns of data would be exported with the PDF.
3. **Click the Turn All Fields Off button .**

4. **Down toward the bottom, select County, Population, and Seniors. These are the only columns that will be included. Click OK.**

5. **In the table of contents, right-click the layer, and click Open Attribute Table. Notice that now you see only the selected columns.**

6. **Close the attribute table.**

Export the map as a geoenabled PDF

1. **On the File menu, click Export Map.**

2. **Set the save location, and select Save as type as PDF. Name the file** seniors.pdf**.**

3. **Under Options, click the Advanced tab. From the Layers and Attributes list, select Export PDF Layers and Feature Attributes.** This option will enable the underlying attribute data to be exported with the map.

4. **Click Save.**

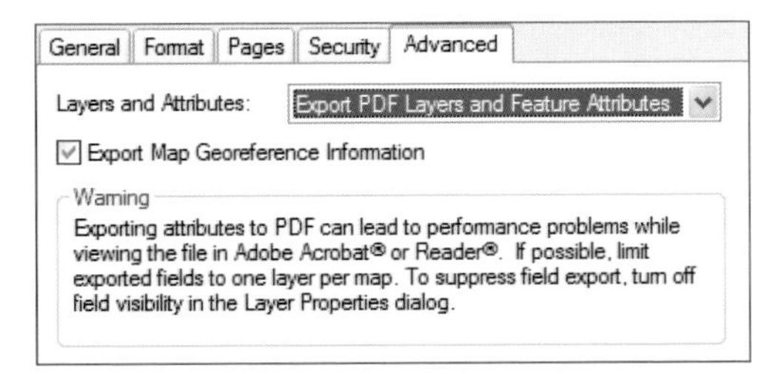

Look at the PDF

NOTE: *There are lots of versions of Adobe Reader, which can be confusing. If you don't see the buttons mentioned in this section, look around in the drop-down menus. The buttons will be somewhere.*

1. **On your desktop, navigate to your save folder, and double-click on the seniors.pdf.** If you have Adobe Reader, it should open using that. If not, Chrome will still display the PDF, but you won't have all the options mentioned next. (If you don't have Adobe Reader, you can download it at www.adobe.com.)

2. **In Adobe Reader, on the navigation menu on the left, click the Layers button.** It looks like sheets of paper stacked on top of each other.

3. **Click the plus sign next to Layers to expand the menu. Click the second plus sign as well.**

4. **Although you have only one layer open, click the button that looks like an eye to turn map elements off and on.** If you had multiple layers, you could turn specific layers off and on to suit your needs.

Access attribute information from within the PDF

There are many versions of Adobe Reader, and the Object Data tool you'll need is tucked into a lot of different spots, depending on the version. In older versions, it's on the Edit menu. In newer versions, you must click Tools, and then click Open on the Measure tool, and the tool will be placed at the top.

1. **In Adobe Reader, click the Object Data tool.**

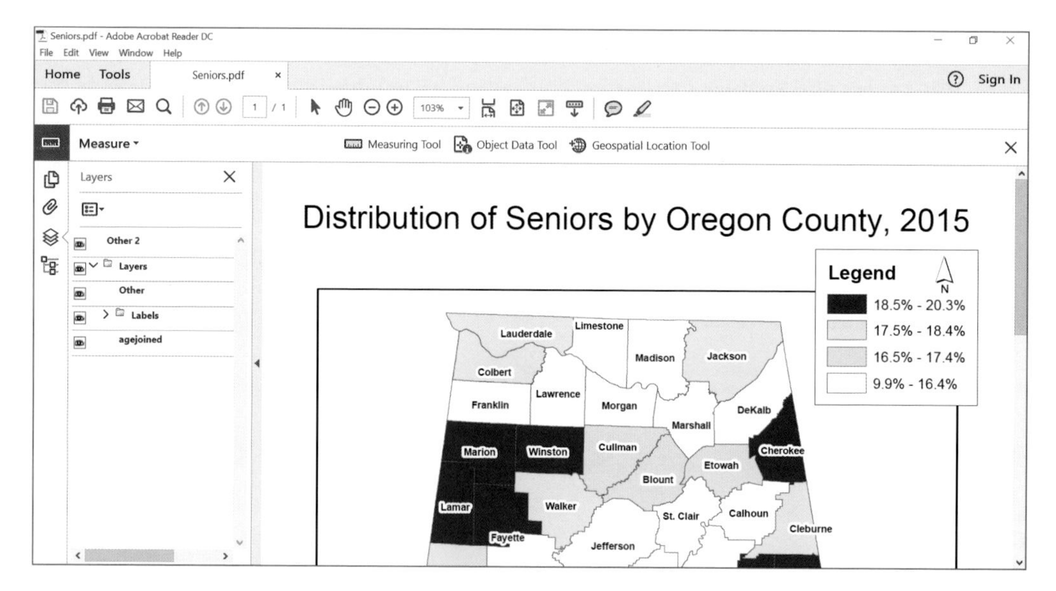

2. **On the map, double-click any county, which will be outlined in red.** Attribute information for the selected county displays on the left at the bottom of the screen.

3. **Close the two boxes above the bottom informational box by dragging up the bar at the bottom center of each box.**

4. **Close Adobe Reader and ArcMap. There's no need to save anything.**

Congratulations on finishing this book! Go out there and do good work with your newly earned GIS skills.

Thanks

Many people helped with the creation of this book. First and foremost, I owe a debt of gratitude to my students, who, over the years, asked all the right questions and shared their ideas and projects with me, which is the basis of this book. Many thanks to Richard and Erez Lufrano for providing the motivation. Thanks to Joan Lufrano, who gave me the idea in the first place. Thanks to Feng Zhang for meticulously checking the steps and making sure non-English speakers wouldn't encounter any roadblocks. Thanks to Esri Press for its continued support.

Source credits

Data credits

Chapter 1

\GIS20\Chapter 1\AlabamaCounties.shp, courtesy of US Census Bureau.

\GIS20\Chapter 1\mystate.mxd, created by the author from US Census Bureau data.

\GIS20\Chapter 1\tl_2016_01_place.shp, courtesy of US Census Bureau.

\GIS20\Chapter 1\tl_2016_01_us_county.shp, courtesy of US Census Bureau.

Chapter 2

\GIS20\Chapter 2\AlabamaCounties.shp, courtesy of US Census Bureau.

\GIS20\Chapter 2\mystate.mxd, created by the author from US Census Bureau data.

\GIS20\Chapter 2\tl_2016_01_place.shp, courtesy of US Census Bureau.

Chapter 3

\GIS20\Chapter 3\AlabamaCounties.shp, courtesy of US Census Bureau.

\GIS20\Chapter 3\countiesprj.shp, created by the author from US Census Bureau data.

Chapter 4

\GIS20\Chapter 4\age.xlsx, created by the author from US Census Bureau data.

\GIS20\Chapter 4\2015 ACS_15_5YR_S0101_with_ann.csv, courtesy of US Census Bureau.

Chapter 5

\GIS20\Chapter 5\age.xlsx, created by the author from US Census Bureau data.

\GIS20\Chapter 5\agejoined.shp, created by the author from US Census Bureau data.

\GIS20\Chapter 5\countiesprj.shp, created by the author from US Census Bureau data.

Chapter 6

\GIS20\Chapter 6\agejoined.shp, created by the author from US Census Bureau data.

\GIS20\Chapter 6\Seniors.mxd, created by the author from US Census Bureau data.

Chapter 7

\GIS20\Chapter 7\agejoined.shp, created by the author from US Census Bureau data.

\GIS20\Chapter 7\minifile.shp, created by the author from US Census Bureau data.

Chapter 8

\GIS20\Chapter 8\agencies.mxd, created by the author from US Census Bureau data.

\GIS20\Chapter 8\Agencies.xls, USADATA courtesy of Dun & Bradstreet.

\GIS20\Chapter 8\Geocoding_Result, created by the author from USADATA® courtesy of Dun & Bradstreet.

\GIS20\Chapter 8\tl_2016_48209_areawater.shp, courtesy of US Census Bureau.

\GIS20\Chapter 8\tl_2016_48209_edges.shp, courtesy of US Census Bureau.

Chapter 9

\GIS20\Chapter 9\agencies.mxd, created by the author from US Census Bureau data.

\GIS20\Chapter 9\tl_2016_48029_areawater.shp, courtesy of US Census Bureau.

\GIS20\Chapter 9\tl_2016_48029_edges.shp, courtesy of US Census Bureau.

Chapter 10

\GIS20\Chapter10\LatLongsPlotted.shp, USADATA courtesy of Dun & Bradstreet.

\GIS20\Chapter10\tl_2016_48029_edges.shp, courtesy of US Census Bureau.

\GIS20\Chapter10\XYData.xls, USADATA courtesy of Dun & Bradstreet.

Chapter 11

\GIS20\Chapter11\newsites.shp, created by the author from US Census Bureau data.

\GIS20\Chapter11\Southwest.shp, created by the author from US Census Bureau data.

\GIS20\Chapter11\States.shp, courtesy of US Census Bureau.

Chapter 12

\GIS20\Chapter12\agejoined.shp, created by the author from US Census Bureau data.

\GIS20\Chapter12\seniors.shp, created by the author from US Census Bureau data.

Chapter 13

\GIS20\Chapter13\seniors.shp, created by the author from US Census Bureau data.

\GIS20\Chapter13\tl_2016_01_place.shp, courtesy of US Census Bureau.

Chapter 14

\GIS20\Chapter14\buffers.shp, created by the author from US Census Bureau data.

\GIS20\Chapter14\circle.shp, created by the author from US Census Bureau data.

\GIS20\Chapter14\divisions.shp, created by the author from US Census Bureau data.

\GIS20\Chapter14\Pacific_NW.shp, created by the author from US Census Bureau data.

\GIS20\Chapter14\States.shp, courtesy of US Census Bureau.

\GIS20\Chapter14\Southwest.shp, created by the author from US Census Bureau data.

\GIS20\Chapter14\tl_2016_48029_edges.shp, courtesy of US Census Bureau.

\GIS20\Chapter14\youthcenters.shp, USADATA courtesy of Dun & Bradstreet.

Chapter 15

\GIS20\Chapter15\gdb.gdb, created by the author from US Census Bureau data.
\GIS20\Chapter15\agejoined.shp, created by the author from US Census Bureau data.
\GIS20\Chapter15\Agencies.xls, USADATA courtesy of Dun & Bradstreet.
\GIS20\Chapter15\Alamo.jpg, courtesy of DigitalGlobe®.
\GIS20\Chapter15\tl_2016_48029_edges.shp, courtesy of US Census Bureau.

Chapter 16

\GIS20\Chapter16\AlabamaCounties.shp, created by the author from US Census Bureau data.
\GIS20\Chapter16\spatialjoin.shp, created by the author from US Census Bureau data.
\GIS20\Chapter16\tl_2016_01_place.shp, courtesy of US Census Bureau.

Chapter 17

\GIS20\Chapter17\Alamo.jpg, courtesy of DigitalGlobe.
\GIS20\Chapter17\tl_2016_48029_edges.shp, courtesy of US Census Bureau.

Chapter 18

\GIS20\Chapter18\agejoined.shp, created by the author from US Census Bureau data.

Chapter 19

\GIS20\Chapter19\agejoined.lpk, created by the author from US Census Bureau data.
\GIS20\Chapter19\agejoined.shp, created by the author from US Census Bureau data.
\GIS20\Chapter19\circle.shp, created by the author from US Census Bureau data.
\GIS20\Chapter19\Clip.gpk, created by the author from US Census Bureau data.
\GIS20\Chapter19\clippedagain.shp, created by the author from US Census Bureau data.
\GIS20\Chapter19\seniors.mpk, created by the author from US Census Bureau data.
\GIS20\Chapter19\Seniors.mxd, created by the author from US Census Bureau data.
\GIS20\Chapter19\States.shp, courtesy of US Census Bureau.

Chapter 20

\GIS20\Chapter20\agejoined.shp, created by the author from US Census Bureau data.
\GIS20\Chapter20\seniors.mpk, created by the author from US Census Bureau data.
\GIS20\Chapter20\Seniors.mxd, created by the author from US Census Bureau data.

Figure credits

Chapter 3

Chapter 3, graphic of UTM zones, courtesy of the USGS.
Chapter 3, graphic of state plane coordinate system, created by the author from files in ArcGIS.

Data license agreement

Downloadable data that accompanies this book is covered by a license agreement that stipulates the terms of use.

Index

Symbols

% symbol, 50

A

accuracy, 32
Add Control Points tool, 141
Add Data button, 6, 8
    datum information, 33
    latitude-longitude points, 96–97
    shapefiles, 6, 68, 110, 114
    shapefiles for reports, 146
    social service agencies, 78
    States.shp, 100
    thematic maps, 58
Add Data tool, 7
addresses, geocoding, 90
address locator (credits required): adding shapefiles, 83–84
    ArcGIS Online via ArcMap, 79
    ArcGIS Trial, 77–78
    changing symbols, 82–83
    geocoding addresses, 80–82
    saving project, 84
    social service agencies, 78–79
    symbolizing layers, 84
address locator (labor required): creating, 85–87
    fixing addresses, 89–90
    geocoding, 88
    geocoding unmatched addresses, 88–89
    opening map, 88
address mapping: overview, 76. *See also* geocoding
Adobe Reader, 167–68
aerial photographs, exporting, 131

aerial photography: adding images to maps, 138
control points, 141–42
Fit to Display, 141
identifying intersections, 139–41
lack of spatial reference, 138
moving between layers, 138
shapefiles, 138
updating georeferencing, 143
View Link table, 143
agencies, displaying, 93
agencies.mxd project, saving, 84
Align Left button, 147
American Factfinder, 40
APA web citation, 66
.apk files, 157
Append tool, 122–24
ArcCatalog, opening, 128–29
ArcGIS, signing in via ArcMap, 79
ArcGIS Online: accessing data, 162
Gallery link, 160–61
location and geographic scale, 161–62
online public account, 160
saving maps, 163
sharing using social media, 163
storage and fees, 166
uploading to, 165
ArcMap: adding files, 37
adding shapefiles, 5–9
basemaps, 166
country column, 47–48
renaming worksheet, 48
signing into ArcGIS, 79
spreadsheet cleanup, 47
symbols, 82
ArcSDE geodatabase, 132
ArcToolbox button, 36, 104
appending shapefiles, 122
arrowhead. *See also* north arrow
Attribute button, 69
attribute queries: erasing, 111
in reports, 149
shapefiles, 110
shapefiles for counties, 111
writing, 110
attributes, reviewing, 92
attribute table: and appending, 124
opening, 146
attribute tables: adding columns, 69–70
editing data, 68
editing outside data, 69
opening, 68
working with multiple, 73

B

basemaps in ArcMap, 166
boundaries. *See* joining boundaries
Browse button, 4, 14, 55
creating shapefiles, 102
deleting columns, 72–73
geocoding addresses, 81
buffering, 115–19
buttons: Add Data, 6, 8, 33, 58, 68, 78, 96, 110, 114, 146
Add Data button, 6, 100
Align Left, 147
ArcToolbox, 36, 104, 122
Attribute, 69
Browse, 4, 14, 55, 72, 81, 102
Catalog, 128
Clear Selected Features, 103, 111
Connect to Folder, 7, 129
Coordinate System, 105
Create Features, 103
Delete Link, 142
Editor Toolbar, 68
Edit Vertices, 104–5
Ellipsis, 59, 71
Export Report to File, 148
Left Alignment, 29
List By Drawing Order, 10
Output Coordinate System, 36
Share, 163, 165
Table Options, 146

Turn All Fields Off, 166
Zoom Whole Page, 25. *See also* tools

C

calculations, making, 70–71
Catalog button, 128
categorical maps: attributes, 92
creating, 92–93
displaying agencies, 93
grouping categories, 94
Census. *See* US Census data download
centroid, explained, 115
Choose Measurement Type tool, 120
city shapefiles, downloading, 3–4
Clear Selected Features button, 103, 111
Clear Selected Features tool, 121
Clip tool, 124–25
color ramp, changing, 59
colors, changing, 62
columns: deleting, 72–73
sorting, 71
columns, adding to attribute tables, 69–70
connecting to folders, 7
Connect to Folder button, 7, 129
control points, adding, 141–42
Coordinate System button, 105
coordinate systems, 32
counties: creating shapefiles, 111
labels, 63
county column, cleaning up, 47–48
county shapefiles: creating, 14–15
downloading, 3–4
Create Features button, 103
Create Features tool, 106

D

data: downloading, 44–45
searching, 42–3
viewing, 12
data tables: adding existing data, 68
attribute tables, 68, 73
calculations, 70–71
columns for attribute tables, 69–70
deleting columns, 72–73
editing data, 69
joining to maps, 54
shapefiles, 68
data view, 24
datum: defined, 32
getting information, 33–34
decimal places, changing, 59
default pointer, 9
Delete Link button, 142
deleting. *See* removing
columns, 72–73
legends, 65
Dissolve tool, 125
distance accuracy, 32
downloading: city shapefiles, 3–4
county shapefiles, 3–4
data, 44–5
data from US Census, 40–46
state shapefiles, 3–4
downloading shapefiles, 2
dragging files, 131

E

editing: constructing shapefiles, 106–7
creating shapefiles, 104–5
data in attribute table, 68
Editor toolbar, 100
features on shapefiles, 103–4
moving polygons and cutting, 101–2
repurposing shapefiles, 102–3
shapefiles, 100
state outline, 100
Editor toolbar, Merge tool, 120
Editor Toolbar button, 68
Edit Vertices button, 104–5
elements, selecting, 9
Ellipsis button, 59, 71
erasing queries, 111. *See also* deleting, removing

Excel: importing into, 45–46
renaming worksheets, 49
Excel files, removing, 56
Excel tables, exporting, 131–32
exporting: aerial photographs, 131
Excel tables, 131–32
reports, 148
selected records, 111
Export Report to File button, 148

F

file geodatabase, creating, 129
files: adding to ArcMap, 37
adding to join, 52
adding to joins, 52
appending, 122–24
dragging, 131
finding, 6–7
merging, 120–21
naming, 15, 49
opening, 131
projecting, 36–37
selecting from US Census Bureau website, 2–3
in shapefiles, 6
unioning, 121–22. *See also* shapefiles
finding files, 6–7
Find tool, aerial photography, 139
FIPS (Federal Information Processing Standard) codes, 7, 46
FIPS columns, double-checking and finding, 52–53
FIPSZone, 34
Fit to margins, 26
folders. *See* save folder
connecting to, 7
fonts: changing, 26
resizing, 29
types, 27
Full Extent tool, 9

G

GCS (geographic coordinate systems), 32
.gdb files, 132
geocoding: addresses, 90
address locator, 77–84
overview, 76. *See also* address locator (credits required), address locator (labor required), address mapping
geodatabases: aerial photographs, 131
ArcCatalog, 128–29
Excel tables, 131–32
file geodatabase, 129
importing shapefiles, 130
overview, 128
types, 132
geoenabled PDFs, publishing, 166–68
geography, selecting, 40–42
geoprocessing packages, 155–57
geoprocessing tools: appending, 122–24
buffers, 118
clipping, 124–25
dissolving, 125
files and symbology, 118–19
merging, 120–21
shapefiles, 118
unioning, 121–22
georeferencing, updating, 143
GeoTagged Photos to Points tool, 96
.gpk files, 155
GPS point mapping: latitude-longitude points, 96
placing points, 96
shapefile of XY events, 96

H

Halo option, 23
hollow colors, 19

I

Identify tool, 9
images, resizing, 25

importing, shapefiles, 130
inserting: legend, 27
    north arrow, 28
    scale bar, 28
    source using text, 29
    title, 26
Intersect tool, 115, 125

J

joining: data tables to maps, 54
    data tables to maps, 54
joining boundaries: attribute table, 135
    shapefiles, 134
    spatial join options, 134–35
    Spatial Join tool, 134
    viewing results, 135
joins: adding files, 52
    creating shapefiles, 55–56
    verifying function, 54–55

L

labels: counties, 63
    fixing, 21–22
    turning on, 20–21
land parcels, unique IDs, 46
landscape orientation, 24
latitude-longitude points, 96–97
layer colors, changing, 18–19
layer packages, 154–55
layers: displaying, 10
    making semitransparent, 20
    moving, 10
    naming, 11
    reordering, 20
    symbolizing, 84
layouts, creating, 24–25, 63
layout view, 22–24
Left Alignment button, 29
legend breaks, changing, 60–61
legends: deleting, 65
    fixing, 63–65
    highest values, 62
    inserting, 27
    organizing, 28
links, deleting, 142
List By Drawing Order button, 10
location queries: creating, 114–15
    shapefiles, 114
locator packages, 157
.lpk files, 154
.lyr files, 50

M

map flow, 30
map fonts, 27
map frame, resizing, 26
map packages: geoprocessing packages, 155–57
    layer packages, 154–55
    saving to desktop, 152–54
map projections. *See* projections
maps, embedding in websites, 163
maps and data tables. *See also* publishing maps
margins, fitting to, 26
.mdb files, 132
Measure tool, 119
merging features, 120–21
moving: and cutting polygons, 101–2
    layers, 10
    shapefiles, 10
.mpk files, 152

N

NAD 1983, 32–33
naming files, 15, 49
north arrow, inserting, 28

O

orientation, changing, 24–25
Output Coordinate System button, 36

P

Package Locator, 157
Pan tool, 9

PDFs, publishing, 166–68
percentage column, 50
percentages, fixing, 59–60
plane projections. *See also* projections
plane projections, looking up, 34–35
pointer, 9
points, placing on maps, 96
Point tool, 106
polygons, moving and cutting, 101–2
portrait orientation, 24
print area, layout view, 25
projecting files, 36–37
projecting shapefiles: adding files to ArcMap, 37
datum information, 33–34
projecting files, 36–37
state plane projections, 34–35. *See also* shapefiles
projections: guide to, 33
problems, 32
projects: saving, 15–16. *See also* saving projects
projects, saving, 15–16, 29, 49
publishing maps: accessing data, 162
ArcGIS Online Gallery, 160–61
ArcGIS Online Public Account, 160
geoenabled PDFs, 166–68
location and geographic scale, 161–62
saving, 163
social media, 163
uploading to ArcGIS Online, 165
web apps, 164–65

Q

queries. *See* attribute queries, location queries

R

reference maps: creating layouts, 24–25
data and layout views, 22
fixing labels, 21–22
inserting titles, 26
layer colors, 18–19
legends, 27
north arrow, 28
reordering layers, 20
saving, 29
scale bars, 28
semitransparent layers, 20
source using text, 28–29
turning on labels, 20–21
removing, shapefiles, 56
removing Excel files, 56
reports: attribute queries, 149
attribute table, 146
creating, 146–48
exporting, 148
shapefiles, 146
resizing: fonts, 29
images, 25
map frames, 26

S

save folder, setting up, 2
saving: projects, 15–16, 30, 49
spreadsheets, 50
saving projects: agencies.mxd, 84
thematic maps, 66
scale bar, inserting, 28
scanned maps, 143
SDE (Spatial Database Engine), 132
.sde files, 132
searching, data, 42–43
selecting: files from US Census Bureau website, 2–3
geography, 40–42
selecting elements, 9
shapefiles: adding, 68, 110, 114, 134
adding to ArcMap, 5–9
cities, 3–4
constructing, 106–7
contents, 12
counties, 3–4
creating, 55, 104–5
creating features, 103–4

creating for counties, 14–15, 111
creating from shapefiles, 102–3
customizing, 12–15
downloading, 2–4
files in, 6
geocoded maps, 83–84
importing, 130
maps and data tables, 12
moving, 10
removing, 56
reports, 146
states, 3–4
XY events, 96–97. *See also* files, projecting shapefiles
Share button, 163, 165
sharing work. *See* map packages
sorting: columns, 71
entries, 12
source, inserting using text, 29
spatial join options, filling in, 134–35
Spatial Join tool, 134
spreadsheet: cleaning up, 47
saving, 49
state outline, editing, 100
state plane versus UTM, 35
state shapefiles, downloading, 3–4
symbolizing layers, 84
symbology, changing, 118–19
symbology, choosing, 62
symbols, changing in ArcMap, 82

T

Table of Contents window, 10–11
Table Options button, 146
Table Options tool, 111
thematic maps: changing color, 62
color ramps, 59
creating, 58
creating layouts, 63
fixing legend, 63
labeling counties, 63
legend breaks, 60–61
percentages, 59–60
placing highest values, 62
saving, 66
titles, 63
titles: inserting, 26
thematic maps, 63
tools: Add Control Points, 141
Add Data, 7
Append, 122–24
Choose Measurement Type, 120
Clear Selected Features, 121
Clip, 124–25
Create Features, 106
deactivating, 9
Dissolve, 125
Find, 139
Full Extent, 9
GeoTagged Photos to Points, 96
Identify, 9
Intersect, 115, 125
Measure, 119
Pan, 9
Point, 106
Spatial Join, 134
Table Options, 111
Union, 121–22
Up One Level, 105
Zoom In, 10
zooming in and out, 8. *See also* buttons
Tools toolbar, 25
Turn All Fields Off button, 166

U

Union tool, 121–22
unmatched addresses, 90
updating, georeferencing, 143
Up One Level tool, 105
US Census Bureau website, APA citation, 66
US Census Bureau website, selecting files, 2–3
US Census data download: American Factfinder, 40
FIPS code, 46

geography, 40–42
importing into Excel, 45–46
performing, 44–45
prepping for ArcMap, 47–49
searching data, 42–43
UTM versus state plane, 35

V

viewing data, 12
View Link table, 142–43
views, data versus layout, 22–24

W

web apps, sharing by, 164–65
WGS84 (World Geodetic System), 32
worksheet, renaming, 49

X

XY events, shapefile, 96–97

Z

zooming in and out, 8
Zoom In tool, 10
ArcGIS Online Gallery, 160–61
shapefiles and Editor toolbar, 100
Zoom Out tool, 160–61
Zoom Whole Page button, 25